Don't Be Evil

RANA FOROOHAR

Don't Be Evil

The Case Against Big Tech

ALLEN LANE
an imprint of
PENGUIN BOOKS

ALLEN LANE

UK | USA | Canada | Ireland | Australia
India | New Zealand | South Africa

Allen Lane is part of the Penguin Random House group of companies
whose addresses can be found at global.penguinrandomhouse.com

First published in the USA by Random House 2019
First Published in Great Britain by Allen Lane 2019
001

Copyright © Rana Foroohar, 2019

The moral right of the author has been asserted

Printed and bound in Great Britain by Clays Ltd, Elcograf S.p.A.

A CIP catalogue record for this book is available from the British Library

HARDBACK ISBN: 978–0–241–40428–7
TRADE PAPERBACK ISBN: 978–0–241–42790–3

www.greenpenguin.co.uk

MIX
Paper from
responsible sources
FSC® C018179

Penguin Random House is committed to a
sustainable future for our business, our readers
and our planet. This book is made from Forest
Stewardship Council® certified paper.

For Alex and Darya

I had worked hard for nearly two years, for the sole purpose of infusing life into an inanimate body. For this I had deprived myself of rest and health. I had desired it with an ardor that far exceeded moderation; but now that I had finished, the beauty of the dream vanished, and breathless horror and disgust filled my heart.

—MARY SHELLEY, *Frankenstein*

Contents

Author's Note

Some books are born out of big, abstract ideas; others start closer to home. My last book, *Makers and Takers,* came out of high-level policy conversations about the financial industry. This book, which examines the economic, political, and cognitive damage wrought by the technology industry over the past twenty years, has a wide lens. But its birth was quite intimate.

It began one afternoon in late April 2017, when I came home from work, opened my credit card bill, and got a shock: over $900 in charges I didn't recognize from Apple's App Store. First, I thought I'd been hacked. After making a few inquiries, however, I discovered that, in fact, my then ten-year-old son had racked up these charges, buying virtual players for the online soccer game he had become fond of.

His devices were summarily confiscated and passwords revoked, needless to say. But right around that time, the larger implications of this incident were beginning to consume my time and attention, albeit in a different way. I was starting a new job, as the global business columnist for the *Financial Times,* the world's largest business newspaper. My mandate was to write a weekly column about the biggest economic stories of the day. And most turned out to involve the Big Tech behemoths of our era, companies like Google, Facebook, Amazon, and, of course, Apple.

It's no secret that the concentration of market power has been rising in numerous industries over the past few decades, a trend that has been linked to everything from growing income inequality to slower economic growth to a surge in political populism. But as I settled into my new role at the *FT* and began digging into the financial data, I discovered something rather shocking—that 80 percent of corporate wealth was now being held by just about 10 percent of companies.[1] And these weren't the firms that owned the most physical assets or commodities; they weren't the GEs or the Toyotas or the ExxonMobils. Rather, they were those that had figured out how to leverage the new "oil" of our economy—information and networks.

Many of these new superstars were technology companies. The tech industry provides the starkest illustration of the rise in monopolistic power in the world today. Ninety percent of the searches conducted everywhere on the planet are performed on a single search engine: Google.[2] Ninety-five percent of all Internet-using adults under the age of thirty are on Facebook (and/or Instagram, which Facebook acquired in 2012).[3] Millennials spend twice as much time on YouTube as they do on all other video streaming services combined.[4] Google and Facebook together receive around 90 percent of the world's new ad spending, and Google's and Apple's operating systems run on all but 1 percent of all cellphones globally.[5] Apple and Microsoft supply 95 percent of the world's desktop operating systems.[6] Amazon takes half of all U.S. e-commerce sales.[7] The list goes on and on. Everything in Big Tech goes big or it doesn't go at all—and the bigger it gets, the more likely it is to go bigger still.

THE WEALTH REAPED by the digital giants has been extraordinary. The market capitalizations of the five so-called FAANG companies—Facebook, Apple, Amazon, Netflix, and Google—now exceed the economy of France. Measured by users, Facebook alone is larger than the world's largest country, China.[8] But as the big have gotten bigger, the rest of the economy has suffered. Over the past two decades, as Big Tech has grown, more than half of all public firms have disappeared.[9] Our economy has become more concentrated, and both business dynamism and entrepreneurship have declined.[10]

As I wrote and reported, raising concerns about all of this in the pages of the *FT,* I felt growing unease over the things I was hearing from a broad swath of people—workers, consumers, parents, and investors—who felt that Big Tech was putting their livelihoods or even their lives (or those of people they loved) in jeopardy. There were the mothers and fathers who struggled with tech-addicted children. The workers who'd lost jobs when their businesses went bankrupt trying to compete with Amazon. The entrepreneur who had his ideas and intellectual property stolen by a competitor and lacked the funds to bring his complaint to court. The homeowner who was denied property insurance because the provider's algorithms determined he was too high a risk. Then there were those who simply felt that the entire tech industry wasn't sharing enough of the wealth pie fairly.

And what a pie it is. Big Tech firms are now the richest and most powerful companies on the face of the planet. The inherent desirability of their various products and platforms, coupled with the network effect, in which more users beget still more, and all the data that they harvest as a result, has allowed them to scale to unimaginable dimensions. They have used their size to crush or absorb competitors, to commandeer the personal data of their users, and—in the case of Google, Facebook, and Amazon—to leverage it for their own benefit in the form of highly targeted advertising. They and other Big Tech firms have also offshored much of their exorbitant profit—according to one estimate done by Credit Suisse in 2019, the top ten companies offshoring the most savings, including Apple, Microsoft, Oracle, Alphabet (the parent company of Google), and Qualcomm, had $600 billion sitting in overseas accounts[11]—circumventing the laws and regulations by which ordinary citizens must abide, but which the largest corporations can legally eschew. Silicon Valley has lobbied hard to preserve the tax loopholes that allow all this, bringing to mind the words of economist Mancur Olson, who warned that a civilization declines when the moneyed interests take over its politics.[12]

Certainly, many public officials I spoke with echoed my concerns. Silicon Valley was, after all, built around government—that is, taxpayer—funded innovation. Everything from GPS mapping to touch screens to the Internet itself came out of research originally done or funded by the U.S. Department of Defense, and was later commercialized by Silicon Valley. Yet unlike many other countries, including a number of thriving free markets such as Finland and Israel, the U.S. taxpayer does not reap a penny of the profits these innovations yield.[13] Instead, these companies were offshoring both profits and labor at the very time that tech titans were asking the government to spend more money on things like educational reform to ensure that the twenty-first-century workforce would be digitally savvy. The consequences are not just economic; by fueling populist discontent with capitalism and liberal democracy, they have high-stakes political ramifications, too.

For someone who'd been tracking the financial industry closely since 2007, the parallels were fascinating. There was a new too-big-to-fail, too-complex-to-manage industry out there, one that had grown like ragweed right under our noses. It had more wealth and a bigger market capitalization than any other industry in history, yet it created fewer and fewer jobs than the behemoths of the past. It was reshaping our economy and labor force in profound ways, turning people into products via the collection and monetization of their personal data, and yet went virtually unregulated. And much like the banking system circa 2008, it was flexing its considerable political and economic muscle to ensure that things stayed that way.

As I began looking more closely at these companies, which were already under fire in the wake of revelations about the 2016 election results, a picture began to form. As we now know, the largest technology platform companies in the world, including Facebook, Google, and Twitter, were exploited by Russian actors to manipulate the results of the U.S. presidential election in favor of Donald J. Trump. These platforms were no longer just places to search for cheap airfare, post vacation photos, or connect with long-lost family and friends. Instead, they had become tools for manipulating geopolitics and swinging the fate of nations—while enriching their executives and shareholders in the process. The innocence of an earlier era was behind us.

That's an important point to remember, because when it comes to the tech industry, it hasn't always been all about the money. In fact, Silicon Valley was heavily influenced by the counterculture movements of the sixties, with many entrepreneurs inspired by a vision of a future in which technology held the power to make the world a better, safer, and more prosperous place for everyone. The "digital utopians" preaching this vision adhered to a strict gospel: that information wanted to be free, and that the Internet would be a democratizing force, leveling the playing field for us all. There was a time when the high priests of the Internet were not cited on *Forbes* list of the richest people on the planet; rather, they were cited across the newly created blogosphere as the creators of Linux and Wikipedia and other open-source platforms, communities built on the assumption that trust and transparency would prevail over greed and profit.

All of which raises the question: How did we get here? How did an industry that had once been scrappy, innovative, and optimistic become, in the span of just a few decades, greedy, insular, and arrogant? How did we get from a world where "information wants to be free" to one in which data exists to be monetized? How did a movement built on the goal of democratizing information come to all but destroy the very fabric of our democracy? And how did its leaders go from tinkering with motherboards in their basements to dominating our political economy?

The answer, as I soon came to believe, is that we reached a tipping point in which the interests of the largest tech firms and the customers and citizens they supposedly served were no longer aligned. Over the past twenty years, Silicon Valley has given us amazing things, from search to social media to portable devices with astounding computing power. We hold today in our pockets more computing intelligence than entire companies had access to just a generation ago. And yet, these modern conveniences have come at a steep price: twitchy technology addiction that saps our time and productivity, the spread of misinformation and hate speech, predatory algorithms targeting the weak and vulnerable, a total loss of personal privacy, and the accumulation of more and more of the country's wealth by a smaller and smaller subset of society.

What's more, all of these problems—while often spoken about in isolation—are intertwined. There was a single, inescapable problem: a business model based largely on keeping people online as long as possible, and monetizing their attention. That's something that many people in the Valley didn't want to acknowledge. The "attention merchants," as Columbia University academic Tim Wu has labeled Big Tech firms, use behavioral persuasion, troves of personal data, and network effects to achieve monopoly power, which ultimately affords them political power, which in turn helps them hold on to their monopolies.

In the past, Facebook, Google, and Amazon have been given regulatory get-out-of-jail-free cards. After all, the logic goes, Google provides its searches for "free." Facebook is "free" to join. And Amazon cuts prices and gives away products for free. Isn't that *good* for consumers? The problem is that "free" isn't actually free. We don't pay for most digital services in dollars—but we do pay dearly, with our data and our attention. *People* are the resource that's being monetized. We think we are the consumers. In fact, we are the product.

OF COURSE, THESE are problems many leaders in Silicon Valley don't much want us to land on. Too many powerful people there remain in a cognitive bubble, reluctant to engage fully and transparently with legitimate public concerns, including safeguarding our data, whether artificial intelligence and automation will take too many jobs, our loss of privacy as our location is tracked on a second-by-second basis by thousands of apps, election manipulation, and even what the shiny devices that permeate every aspect of our lives today are doing to our brains. When I ask most techies about these concerns, reactions tend to range from defensive to naïve to clueless or, the worst of all, a patronizing smile or exasperated look that says "You're not a tech insider, and thus you just don't get it."

But it may be the tech executives themselves who don't get it. As John Battelle, who helped launch *Wired* magazine, once put it to me, "The tech community doesn't have a good perspective on itself. We aren't humanists or philosophers. We are engineers. To Google and Facebook, people are algorithms."[14]

It all seems too familiar. I am old enough to have lived through one big boom-and-bust tech cycle, having worked for a high-tech incubator in London from 1999 to 2000, an experience I'll describe in a later chapter of this book. Then, as now, the industry was talking mainly to itself. The hubris we see today has reached levels we haven't seen since the years leading up to the dot-com collapse—only this time around, it's more pernicious, given that companies like Amazon and Apple have become mainstays in just about every household in America. Like the big Wall Street banks, they hold vast amounts of money and power and even greater troves of data. Yet, unlike Goldman Sachs chief executive Lloyd Blankfein, they are not joking when they claim to be doing God's work. Attend any tech conference and you'll quickly learn that many in Silicon Valley still subscribe to the notion that they have made the world more free and open, despite plenty of evidence to the contrary.

The Valley has clearly moved away from its hippie, entrepreneurial roots. Big Tech chief executives are as rapaciously capitalist as any financier, but often with an added libertarian bent. Theirs is a worldview in which anything and everything—government, politics, civic society, and law—can and should be disrupted. As Big

Tech critic Jonathan Taplin once put it to me, "Demos—society itself—is often viewed as being 'in the way.' "[15]

So why haven't our political leaders imposed any sensible regulations to hold such instincts in check? Follow the money. Not for nothing that Big Tech now vies with Wall Street and Big Pharma as one of the top spenders on political lobbying. Just as, in the years before the 2008 financial crisis, the world's top bankers dispatched surrogates to Washington, London, and Brussels to live among and lobby the legislators in charge of regulating them, so Silicon Valley faces have become the most familiar ones in these capitals over the past decade—with Google dispatching so many emissaries to Washington that they needed office space as large as the White House to hold them all.[16]

But despite the efforts of scores of Silicon Valley lobbyists and PR teams, the public worries about the economic and social effects of technology, and those worries are not going away.[17] In fact, they are increasing, as the technology itself spreads more deeply into our economy, politics, and culture. Big Tech has become the new Wall Street, and as such, is the prime target for a populist backlash in a world increasingly bifurcated, economically and socially.

The changes Big Tech has wrought have become one of the most pressing economic issues of our time. Harvard Business School professor emerita Shoshana Zuboff and other scholars have decried the rise of "surveillance capitalism," which is, as Zuboff defines it, "a new economic order that claims human experience as free raw material for hidden commercial practices of extraction, prediction and sales," as well as "a parasitic economic logic in which the production of goods and services is subordinated to a new global architecture of behavioral modification" via digital surveillance technologies.[18] She believes (and I would agree) that surveillance capitalism represents a significant threat to our economic and political systems, as well as a potential instrument for social control.[19] I've also come to believe that curbing Silicon Valley's nefarious side effects will become "the signature economic issue [for lawmakers] over the next five years, especially as automation increases and they make investments into other areas of the economy," as one staffer for an influential senior Democratic senator has put it to me.

Yet this isn't just a story for the business pages. In fact, Big Tech is at the center of nearly every story in the news today, ranking second only to stories about Donald Trump in the press. Yet while the president will leave us eventually, Big Tech is forever, transforming our existence a little more every day, as the technology itself spreads more deeply into our economy, politics, and culture. It's an alchemy that is just beginning. As amazing as the changes of the past twenty years have been, they are only the first stages of a multi-decade transition to a digital economy that will rival the industrial revolution in terms of transformative power. And by the time it is complete, the consequences are likely to be even more sweeping, changing the nature of liberal democracy, of capitalism, and even of humanity itself.

What Big Tech is doing is, in a word, big. And while I've been critical of many aspects of this digital transformation, there is no denying the tremendous upside as well. Silicon Valley has been the single greatest creator of corporate wealth in history. It has connected the world, helped spark revolutions against oppressive governments (even as it has also facilitated repression), and created entirely new paradigms for invention and innovation. Platform technologies allow many of us to work remotely, maintain distant relationships, develop new talents, market our businesses, and share our views, our creative expression, and/or our products with a global audience. Big Tech has given us the tools to call up a variety of goods and services—from transportation to food to medical treatment—on demand, and generally live in a way that is more convenient and efficient than ever before.

In these and many other senses, the digital revolution is a miraculous and welcome development. But in order to ultimately reap the benefits of technology in a broad way, we need a level playing field, so that the next generation of innovators is allowed to thrive.

We don't yet live in that world. Big Tech has reshaped labor markets, exacerbated income inequality, and pushed us into filter bubbles in which we get only the information that confirms the opinions we already have. But it hasn't provided solutions for these problems. Instead of enlightening us, it is narrowing our view; instead of bringing us together, it is tearing us apart.

With each buzz and beep of our phones, each automatically downloaded video, each new contact popping up in our digital networks, we get just a glimmer of a vast new world that is, frankly, beyond most people's understanding, a bizarre land of information and misinformation, of trends and tweets, and of high-speed surveillance technology that has become the new normal. Just think: Russian election-hacking; hate-mongering Twitter feeds; identity theft; big data; fake news; online scams; digital addiction; self-driving car crashes; the rise of the robots; creepy facial recognition technology; Alexa eavesdropping on our every conversation; algorithms that watch us work, play, and sleep; and companies and governments that control them. The list of technology-driven social disruption is endless—and all of it has appeared in just the past few years. Individually, each item is just a speck in the eye, but collectively it makes for a sleet storm, a freezing whiteout that yields a foggy numbness, the anxious haze of the modern age.

The issue is that periods of great technological change are also characterized by great disruption, which needs to be managed for the sake of society as a whole. Otherwise, you end up with events like the religious wars of the sixteenth and seventeenth centuries, which, as historian Niall Ferguson has outlined in his book *The Square and the Tower*, might not have happened without the advent of major new technologies like the printing press, which eventually brought with it the Age of Enlightenment, but not before it upset old orders in the same way that the Internet and social media have upended society today.[20]

No one can hold back technology—nor should they. But disruption can and should be better managed than it has been in the past.

We have the tools to do so. The challenge for us today is figuring out how to put boundaries around a technology industry that has become more powerful than many individual countries. If we can create a framework for fostering innovation and sharing the prosperity in a much broader way, while also protecting people from the dark side of digital technologies, then the next few decades could be a golden era of global growth.

This book is an attempt to shine a light on the things about Big Tech that should worry us, and what we can do to fix them. I hope that it will serve as a wake-up call, not just for executives and policy makers but for anyone who believes in a future in which the benefits of innovation and progress outweigh the costs to individuals and to society. It's in everyone's interest to believe that we can create that kind of future. Because as we've come to understand all too clearly over the past few years, once people stop believing that a system is good for them, the system falls apart.

Don't Be Evil

CHAPTER 1

A Summary of the Case

"Don't be evil" is the famous first line of Google's original Code of Conduct, what seems today like a quaint relic of the company's early days, when the crayon colors of the Google logo still conveyed the cheerful, idealistic spirit of the enterprise. How long ago that feels. Of course, it would be unfair to accuse Google of being actively evil. But evil is as evil does, and some of the things that Google and other Big Tech firms have done in recent years have not been very nice.

When Larry Page and Sergey Brin first dreamed up the idea for Google as Stanford graduate students, they probably didn't imagine that the shiny apple of knowledge that was their search engine would ever get anyone expelled from paradise (as many Google executives have been over a variety of scandals in recent years). Nor could they have predicted the many embarrassments that would emanate from the Googleplex: Google doctoring its algorithms in ways that would deep-six rivals off the crucial first page of its search results. Google's YouTube hosting instructional videos on how to build a bomb. Google selling ads to Russian agents, granting them use of the platform to spread misinformation and manipulate the 2016 U.S. presidential election. Google working on a potential search engine for China—one that would be compliant with the regime's efforts to censor unwelcome results. Former Google CEO Eric Schmidt leaving his position as executive chairman of Google's parent company, Alphabet, a few months after *The New York Times* revealed he'd been unduly influencing antitrust policy work at a think tank that both his family foundation and Google itself supported, going so far as to push for the firing of a policy analyst who dared to speculate about whether Google might be engaging in anticompetitive practices (something that Schmidt has denied). In May 2019, Schmidt announced he would be stepping down from the Alphabet board as well.[1]

All of this may not exactly be evil, but it certainly is worrisome.

Google's true sin, like that of many Silicon Valley behemoths, may simply be hubris. The company's top brass always wanted it to be big enough to set its own rules, and that has been its downfall, just as it has been for so many Big Tech firms. But this is not a book about Google alone. It is a book about how today's most powerful companies are bifurcating our economy, corrupting our political process, and fogging our minds. While Google will often stand as the poster child for the industry more generally, this book will also cover the other four FAANGs—Facebook, Apple, Amazon, and Netflix—as well as a number of additional platform giants, like Uber, that have come to dominate their respective spaces in the technology industry. I'll also touch on the ways that a variety of older companies, from IBM to GM, are evolving in response to these new challengers. And I will look at the rise of a new generation of Chinese tech giants that is going where even the FAANGs don't dare.

While there are plenty of companies both in Silicon Valley and elsewhere that illustrate the upsides and the downsides of digital transformation, the big technology platform firms have been the chief beneficiaries of the epic digital transformation we're undergoing. They have replaced the industrialism of the nineteenth and twentieth centuries with the information-based economy that has come to define the twenty-first.

The implications are myriad, and I will track many of them,

often via the Google narrative, which has been the marker for larger industry-wide shifts. Google has, after all, been the pioneer of big data, targeted advertising, and the type of surveillance capitalism that this book will cover. It was following the "move fast and break things" ethos long before Facebook.[2]

I've been following the company for over twenty years, and I first encountered the celebrated Google founders, Page and Brin, not in the Valley, but in Davos, the Swiss gathering spot of the global power elite, where they'd taken over a small chalet to meet with a select group of media.[3] The year was 2007. The company had just purchased YouTube a few months back, and it seemed eager to convince skeptical journalists that this acquisition wasn't yet another death blow to copyright, paid content creation, and the viability of the news publications for which we worked.

Unlike the buttoned-up consulting types from McKinsey and BCG, or the suited executives from the old guard multinational corporations that roamed the promenades of Davos, their tasseled loafers slipping on the icy paths, the Googlers were the cool bunch. They wore fashionable sneakers, and their chalet was sleek, white, and stark, with giant cubes masquerading as chairs in a space that looked as though it had been repurposed that morning by designers flown in from the Valley. In fact, it may have been, and if so, Google wouldn't have been alone in such excess. I remember attending a party once in Davos, hosted by Napster founder and former Facebook president Sean Parker, that featured giant taxidermy bears and a musical performance by John Legend.

Back in the Google chalet, Brin and Page projected a youthful earnestness as they explained the company's involvement in authoritarian China, and insisted they'd never be like Microsoft, which was considered the corporate bully and monopolist of the time. What about the future of news, we wanted to know. After admitting that Page read only free news online whereas Brin often bought the Sunday *New York Times* in print ("It's nice!" he said, cheerfully), the duo affirmed exactly what we journalists wanted to hear:

Google, they assured us, would never threaten our livelihoods. Yes, advertisers were indeed migrating en masse from our publications to the Web, where they could target consumers with a level of precision that the print world could barely imagine. But not to worry. Google would generously retool our business model so we, too, could thrive in the new digital world.

I was much younger then, and not yet the (admittedly) cynical business journalist that I have become, and yet I still listened to that happy "future of news" lecture with some skepticism. Whether Google actually intended to develop some brilliant new revenue model or not, what alarmed me was that none of us were asking a far more important question. Sitting toward the back of the room, somewhat conscious of my relatively junior status, I hesitated, waiting until the final moments of the meeting before raising my hand.

"Excuse me," I said. "We're talking about all this like journalism is the only thing that matters, but isn't this really about . . . democracy?" If newspapers and magazines are all driven out of business by Google or companies like it, I asked, how are people going to find out what's going on?

Larry Page looked at me with an odd expression, as if he was surprised that someone should be asking such a naïve question. "Oh, yes. We've got a lot of people thinking about that."

Not to worry, his tone seemed to say. Google had the engineers working on that "democracy" problem. Next question?

Well, it turns out that we did have to worry about democracy, and since November 2016, we have had to worry about it a lot more. And it's impossible to ignore the obvious: As tech firms have become inexorably more powerful, our democracy has become more precarious. Newspapers and magazines have been hollowed out by Google and Facebook, which in 2018 together took 60 percent of the Internet advertising market.[4] This is a key reason for the shuttering of some 1,800 newspapers between 2004 and 2018, a process that has left 200 counties with no paper at all,[5] restricting the supply of reliable information that is the oxygen of democracy.

And given that digital advertising surpassed TV ads in 2017, it's clear that TV news will be the next to go.[6] While cable news may have gotten a "Trump bump" in recent years, the longer term trend line is clear—TV will ultimately be disintermediated by Big Tech just the way print media has been.

But the trouble with Big Tech isn't just an economic and business issue; it has political and cognitive implications as well. Often, these trends are written about in isolation, but in fact they are deeply intertwined. In this book, my goal is to connect the dots—to tell the whole story, which is far bigger than the sum of its parts.

Things Fall Apart: The Political Impact of Big Tech

After it was revealed that the largest technology platforms in the world were exploited by Russian state actors and their private proxies to swing the 2016 U.S. presidential election, it was Facebook, not Google, who took most of the heat. CEO Mark Zuckerberg insistently denied the possibility that nefarious foreign actors could have hacked the platform, which, of course, is exactly what was revealed to have happened. As *The New York Times* later reported, both he and COO Sheryl Sandberg had enlisted a shadowy right-wing PR firm that used underhanded techniques to discredit the Big Tech critic and financier George Soros.

But Google was only marginally more responsive to those first signs of election manipulation in the wake of 2016, and it turned out to have played a major role as well. Its subsidiary YouTube was a host to much of the pre-election hate that was stirred up by actors both abroad (including the same Russian agents that were active on Facebook) and at home.[7]

The 2016 election, Brexit, and the continued role that Russia plays in online disinformation underscore the fact that the very cohesion of society is at stake in this new digital revolution. We are experiencing a crisis of trust in this country; we've lost faith in our institutions, our leaders, and the very systems by which society is governed. As tempting as it might be to point a finger straight at the White House, this is not all about the current administration. For one, research shows that the declining trust in liberal democracy has coincided with the rise of social media.[8] Part of this has to do with the fake news problem—which academic studies have found is 70 percent more likely to be shared than real news.[9] But the fall in trust also has to do with a sense that the game is rigged, and that there is now an even wider social and economic chasm dividing the haves and have-nots, a divide created not just by Wall Street, but by Silicon Valley, too.[10] In 2008, Washington bailed out the largest and most powerful banks and left ordinary homeowners to take losses. We can argue about the economic rationale for this, but the political result was the emergence of a narrative that the system had been captured by a small group of rich and powerful people. It drove voters on both ends of the spectrum away from the Republican and Democratic centers as a result.

Now, just as the public fury at Wall Street after the 2008 crisis contributed to the populist backlash that led to Donald Trump, the sense that Silicon Valley is building robots instead of factories, and creating paper billionaires instead of jobs, is now fueling extremism on both ends of the political spectrum: from the rise of fascism among white men in red states, to socialism among angry young millennials in the blue states (feelings that are, of course, aired and fanned on the very technology platforms that have helped to fuel them). When you stop to think about it, it's not so surprising that a growing number of experts believe that it was tech-based disruption as much as trade that pushed the American Rust Belt toward Donald Trump.[11]

There is no question that the tech sector has spawned incredible economic bifurcation. A 2016 report by the Economic Innovation Group revealed that a mere 75 of America's 3,000-plus counties make up 50 percent of all new job growth. These are the places where Big Tech looms large: San Francisco, Austin, Palo Alto, and so on. The cities where the large tech firms locate create wealth, but often become walled gardens.[12] Witness the protests over housing bubbles in San Francisco that have left even the middle class unable to afford homes.

Then there's the fact that election manipulation via platform technology continues to be a huge problem around the world, with Google and Facebook being used to oppress entire populations or even support genocide and murder in countries from Myanmar to Cameroon.[13] There are some who believe technology is making us more vulnerable to fascism.[14] This is one of the reasons that financier George Soros, founder of the Open Society, has now made the study of Big Tech a key area of his philanthropic work.

Born in Hungary, Soros is acutely sensitive to the political implications of this technical revolution, seeing in it the potential for an authoritarian state to harvest our private data and put the knowledge to nefarious uses of the sort predicted in George Orwell's *1984*. In a speech at Davos in January 2018, he noted that Big Tech was divesting people of their autonomy, explaining that "it takes a real effort to assert and defend what John Stuart Mill called 'the freedom of mind.' There is a possibility," Soros said, "that once lost, people who grow up in the digital age will have difficulty in regaining it." He feared the risk of "alliances between authoritarian states and these large, data-rich IT monopolies that would bring together nascent systems of corporate surveillance with an already developed system of state-sponsored surveillance."[15]

He's right to be fearful. China has its own FAANGs, known as the "BATs"—Baidu, Alibaba, and Tencent—that routinely monitor the Chinese people in "smart cities," a deceptively innocent moniker for 24/7 surveillance areas that are wired up with sensors (in fact, Soros gave his 2019 Davos speech on the dangers posed by the Chinese surveillance state).[16] And the technology that powers these cities, it's worth noting, is produced and installed not only by Chinese firms such as Huawei, but also by American companies like Cisco. The resulting information is, of course, part of the Chinese government's own efforts to move ahead in areas like artificial intelligence that depend on massive amounts of data, or used in the Middle Kingdom's creepy system of "social credits," in which citizens are monitored and given scores that can influence everything from their ability to get loans to where they can live. What isn't garnered from Chinese companies has been taken via partnerships with companies like Facebook (which was, in 2018, revealed to be allowing Huawei and other Chinese firms access to users' nonpublic data).[17]

All of which makes it particularly rich that some Big Tech firms have responded to the growing public concern about privacy and anticompetitive business practices by playing to a long-standing American fear: It's us versus China. Companies like Google and Facebook are increasingly trying to portray themselves to regulators and politicians as national champions, fighting to preserve America's first-place standing in a video-game-like, winner-take-all battle for the future against the evil Middle Kingdom. In the spring of 2018, when Mark Zuckerberg was grilled in front of the U.S. Senate about his company's involvement in election manipulation, an Associated Press reporter managed to take a picture of Zuckerberg's notes, which revealed that if he was asked about Facebook's monopoly power, he had planned to answer that if the company were broken up, America would be at a competitive disadvantage against Chinese tech giants.

As congressional staffers and politicos in Washington have told me, Google has played the national security card, too, quietly using the "U.S. versus China" argument to push back against proposed antitrust action. Yet Google also has a research facility in Beijing, and has contemplated starting a censored version of its search engine to comply with local rules (something that has been put "on hold," as one PR representative put it to me, following an internal revolt among its own engineers, as well as political pushback from the White House and Congress).[18]

Apple doesn't seem to have many qualms about China's "local rules," either. The company may have been protective of user data in the United States, refusing to help the FBI break in to a locked iPhone during investigations of the 2015 San Bernardino terrorist attack, but in China, things are different. When Beijing forced the company to move all of its iCloud data centers for Chinese customers to the mainland, where they would be run by a local company that doesn't need to comply with U.S. laws about data protection, Apple quickly acquiesced, showing that there are limits to its philosophy of preserving civil liberties when there are true threats to its business model in key markets.[19] Even Netflix, which is in some ways the Teflon FAANG, one that comes in for less criticism because of a subscription business model focused on less sensitive data about our entertainment preferences, has bowed to foreign censors. In early January 2019, it emerged that Netflix had pulled an episode of its popular comedy show *Patriot Act* in Saudi Arabia, after government officials complained about one of the actors on the show criticizing Crown Prince Mohammed bin Salman for his role in the murder of Saudi dissident Jamal Khashoggi and for the Saudi war atrocities in Yemen.[20]

Meanwhile, Big Tech is taking on the role of Big Brother right here in the United States, working with local, state, and national authorities to create what is starting to look a lot like a surveillance nation. Amazon sells facial recognition technology to the police. Palantir, the big data firm cofounded by PayPal entrepreneur Peter Thiel, works with the LAPD to target citizens in an alarming manner that might have been drawn from the dystopian thriller *Minority Report*.[21,22] What else that data might be used for is anyone's guess; the clandestine nature of it all makes it nearly impossible to track. But the result is that, little by little, American democracy has ceded a bit more ground to Big Tech.

REGULATORS ARE FINALLY beginning to turn their attention to these issues. In the summer of 2019, as this book goes to press, Google, Facebook, Amazon, and Apple are being investigated by the Department of Justice and the Federal Trade Commission. The

House of Representatives antitrust subcommittee is taking action, too, with plans for months of hearings on Big Tech.[23] But I doubt that the problems will be resolved in time for the 2020 elections—if at all. Despite their professed (and politicized) outrage about Google and Facebook allegedly manipulating their algorithms in favor of liberal politicians, most Republicans are reluctant to touch the issue in a serious way, because to do so would question the legitimacy of the Trump presidency, given the Russian election meddling on his behalf via the same platforms.

Liberals, on the other hand, are divided in their attitudes toward Big Tech. The corporate wing of the party, made up of representatives such as Senator Chuck Schumer of New York, believes in "self-regulation" for Silicon Valley, just as he does for the big banks of his home state. It's telling that Schumer was one of the politicians Facebook tapped in its efforts to limit the fallout over its involvement in election manipulation; and Schumer was only too happy to comply, advising colleagues like the prominent Facebook critic Senator Mark Warner to tone down their criticism of the company (coincidentally or not, Schumer's daughter works at Facebook).[24] The progressive wing is more inclined to take on Silicon Valley (as are, it should be said, some free-market conservatives who don't ally with Trump). And a number of 2020 Democratic candidates have made it a key platform issue. But making changes to the industry will be complicated and require a retooling of many diverse rules and regulations that are supported (or opposed) by a jumble of disparate interest groups.

Meanwhile, the titans of Big Tech—who are often accused of being disproportionately liberal (they are really more libertarian)—are busy throwing support to whichever party will best serve their interests. Former Google CEO Eric Schmidt, for example, gives to Democrats and Republicans, is friendly with the Trump administration, and sat on the Department of Defense Innovation Board under both the Obama and Trump presidencies. Schmidt was also a key adviser in digital efforts for both the Obama and Hillary Clinton campaigns, using Google's might to help the former get elected, and exerting policy influence afterward that is worrisome, to say the least.[25]

While this obviously isn't problematic in the same way that allowing the Trump campaign to spread racist dog whistles and fake news during the 2016 elections was, it underscores the point that these companies hold undue influence over our political system as a whole, in ways that undermine public trust.[26] Schmidt is certainly not alone in playing both sides of the political fence. Take a look at the first meeting of Silicon Valley's tech titans with Donald Trump in 2017, and you'll see Sheryl Sandberg, Tim Cook, and many other avowed Democrats leaning in to the president, literally. Despite Amazon CEO Jeff Bezos's ownership of *The Washington Post*, which is often critical of the president, Amazon pushed its facial recognition technology to ICE, the U.S. Department of Homeland Security's Immigration and Customs Enforcement division—the very one that was keeping children in cages at the Mexican border.[27]

Most Democrats and an increasing number of Republicans have been bought out by Big Tech's extensive lobbying. Silicon Valley is onto a good thing, and, naturally enough, they want to keep it going—which is why they've been silently upping their lobbying presence in Washington, both overtly and covertly. If you combine IT, electronics, and platform technologies, Big Tech is now the second largest lobbying group in our nation's capital, right behind Big Pharma, with Google's parent company Alphabet frequently weighing in as the single largest individual corporate lobbyist in Washington.[28]

Google emerged as the most influential corporate lobbyist—and the one to get more face time from the White House than any other corporate entity—during Barack Obama's second term, just as Big Tech was emerging as what criminal investigators term a "subject of interest." That's when Google, Facebook, and other Big Tech firms began to blanket an unlikely assortment of interest groups with money. The American Library Association, the American Association of People with Disabilities, the National Hispanic Media Coalition, and the Center for American Progress, for example, may not seem like natural allies of the tech revolution, but they have supported some of the regulatory loopholes that Big Tech firms enjoy, including rules that shield them from liability for what users say and do online.[29]

These groups might have reason to object to these tech behemoths on various policy issues, but the heavy donations from their Silicon Valley benefactors often garner tacit support, and sometimes outright endorsement. The ALA, for example, unlike many other groups that represent authors or publishers,[30] supported Google in its fight for the right to scan all the world's books,[31] and while it's true that librarians generally support free speech and want books to be widely accessible, it's also true that Google gives the ALA money and has worked closely with them on various indexing and coding projects. Google has even wormed its way into academia, funding numerous research projects that deal with high-tech issues, and in turn winning favorable commentary from academics who might otherwise be skeptical.[32] In reporting on these issues, I've found it quite difficult to locate completely independent voices on the topics—most experts are funded in some way by either Big Tech firms or their corporate opponents, which goes to show just how thoroughly monied interests have captured the civic debate in the United States. Technologists want to have conversations about economic, political, and social issues on their own terms, or not at all.

The bottom line is that these companies have manipulated the system to ensure that they can continue to operate freely, without the burden of pesky government intervention. The result is that they all too often exist in a universe of their own, not just outside of national borders, but somehow transcending borders altogether. It is in this spirit that Palantir's Peter Thiel and other powerful tech entrepreneurs and investors have suggested that California secede from the Union; Thiel once funded a plan for a network of floating islands that would operate outside of U.S. government jurisdiction, while he and other tech billionaires maintain hideaways in New Zealand.

IN THE MEANTIME, Big Tech itself—like Big Finance before it—has controlled the narrative, using complexity to obfuscate. I cannot tell you how many conversations I have had with fast-talking technologists who try to throw as much jargon against the wall as possible to see what sticks. Yet the simplest questions are often the ones they have the most trouble with. I continue to await a clear answer to the fundamental questions: "Are you playing by the same rules as everyone else? And if not, why not?"

Silicon Valley has always had a core Ayn Rand libertarianism underneath its hippie patina: It justifies their sense of freedom from any costly social responsibility for the downsides of their products and services. As Jonathan Taplin, Jaron Lanier, and other Silicon Valley critics have written, the tech titans may tend to vote left, but the strong libertarian bias in digital culture cuts right. Theirs is an eighties-style "Greed is good" ethos overlaid with the contempt of a youthful generation of CEOs who've never seen government do anything much more ambitious than cut taxes. All of this has resulted in a self-interested and shortsighted "disrupt everything" mentality. It's much easier, of course, to break things than to fix them.

The New Monopolists: Big Tech and
Its Economic Implications

In my nearly three decades of business journalism, I've learned one investigative rule: Follow the money. Big Tech has more of it than any other industry today, and while their meticulous product designs, aggressive marketing, and massive economies of scale have certainly been key drivers of this wealth, Silicon Valley's riches are also a product of a more fundamental economic shift: from an economy based on widgets (and the servicing of widgets) to one based on bits and bytes. Big Tech is redefining what is real and what is of value in our economy, and nothing is more valuable to these companies than our personal data, acquired invisibly from virtually every keystroke we make online, as well as from an increasing number of the moves we make in the physical world. (If you have an Android phone, it knows where you are right now; if you have sensors in household products, they can track things, too.)[33]

When powerful tech firms keep us glued to our devices, it's not really our minds they're after, but rather the data that makes up our consumer profile—a combination of our age, location, marital status, interests, background, education level, political leanings, purchase history, and much more. They then sell this data to third-party marketers, who may in turn sell it to any number of others that want to reach you, from retailers to election manipulators in Russia. It can be deployed in hyper-targeted ads or agglomerated to provide super-detailed forecasting of a variety of social and commercial trends that are of incalculable value to their acquirers.

Such data is the oil of the information age, and it fuels the growth of those companies that can run on it—which is, nowadays, almost every company in almost every industry. This is a very important point—while the problems I'm outlining in this book (loss of privacy, corporate monopoly power, the decline of liberal democracy, and so on) are often best illustrated by the FAANGs, they certainly don't end with them. It's telling that Cambridge Analytica, the British political firm employed by the Trump campaign in the 2016 elections, leveraged information garnered not just from Facebook to create voter profiles, but from dozens of other sources as well, including educational institutions and church groups;[34] in fact, you could argue that the tech companies are simply the canaries in the coal mine for what will eventually become a much larger shift toward a surveillance capitalism system in which businesses and organizations of all stripes will take part. Just as the businesses that figured out how to use mechanized equipment in the industrial age were the ones to thrive, those who are able to make use of this data do the equivalent in our time. And Google and Facebook have figured out how to use all these data points to target ads with the granular precision of a drone strike on an ISIS commander emerging for a cigarette from a bunker somewhere in Syria at 3:13 P.M.

So far, this data has been obtained via computers and mobile devices. But with the rise of personal digital assistants like Amazon's Alexa, Google's Home Mini, and Apple's Siri—now in a third of American homes, with triple-digit sales growth a year—the human voice is the new gold. While reports of Alexa and Siri "listening in" on conversations and phone calls are disputed, there is no question that they can hear every word you say—and from there, it is a short step to them using that knowledge to direct your purchasing decisions. It isn't much of a longer step to see the political implications— already some researchers worry that digital assistants will become even more powerful tools than social media for election manipulation.

Certainly, none of us will be unaffected. Consider that homeowner who was denied insurance, an example which is by no means singular. Since its inception, the insurance business has been based on risk pooling: Total up the cost of insuring a particular group of homes, cars, and lives, and then divide it up evenly, property by property. In the data age, insurance groups will be able to draw information from tracking devices in your car or sensors—like a "smart" thermostat or smoke detector or security camera (perhaps made by Nest Labs, the leader in smart home products, which is owned by Google)—embedded in your house and then use it to price a policy exclusively for you based on your habits and personal style. You might be rewarded for putting a new plumbing system into your own old house (the sensors will measure how well it works), or stopping judiciously at yellow lights. Nice, right?

Here's what's not so nice. You take a price hit when sensors detect your sixteen-year-old puffing weed in his bedroom (smoke detectors will relay the message to your insurer in real time) or if you fail to shovel the snow off the front stoop before it ices up (sensors will record when and if you did and convey that information to the insurance company, limiting their own risk of liability if a passerby slips). You might be given the chance to opt out of all this surveillance, but the insurance company won't make it easy. As it is, when you go onto platforms like Facebook or Google, you can't say no to the surveillance without forfeiting your rights to use many of the services altogether.

It's easy to see how this level of micro-targeting could impact the weakest and more vulnerable; Google, for example, took years before it stopped allowing payday lenders to advertise on the platform, where the personal information they could leverage made it all too easy for them to target vulnerable borrowers.[35] Likewise, the insurance example I've outlined above, which disrupts risk sharing among a larger population and leaves the individual on their own, could ultimately result in an uninsurable underclass, left at the mercy of subprime lenders or the state. Which brings up another dirty secret of the digital age—the fact that the government may well end up being the insurance provider of last resort, placing the burden of insuring those deemed high risk by private companies on the taxpayers.[36]

But that's just the start of it. It's not only Facebook and Google that are collecting data on everyone and leveraging the power of that data to favor the largest and most powerful. Big Tech has shown others the way, and now a vast variety of businesses are developing their own data-mining techniques to get in on this bonanza, spreading tentacles throughout the economy. Data brokers such as credit bureaus, along with healthcare data firms and credit card companies, collect and sell all sorts of sensitive personal user data to other businesses and organizations that do not have the scale to collect it themselves. These include retailers, banks, mortgage lenders, colleges, universities, charities, and—as if we could forget—political campaigns.[37]

Turn on your phone, and you are opening up a world of apps

that track where you are and what you are doing at every second of the day. These apps alone represent a $21 billion industry of snooping, and it's not only the largest tech companies that benefit (though they certainly do; Google's Android system has 1,200 apps that do such tracking), but a host of companies that you probably don't even think about, from Goldman Sachs to the Weather Channel.[38] And that's just the consumer side of things. The old commercial Internet is shifting to an industrial "Internet of things" that will push data harvesting out into the physical world—into design firms, manufacturing plants, insurance companies, financial houses, hospitals, schools, and even our homes.

Name any successful company: Starbucks, Johnson & Johnson, Goldman Sachs . . . and it's likely that successful data mining plays an important role in their business strategy. Real estate companies use a variety of AI applications to mine the data of potential buyers and sellers, even automating the process of home flipping.[39] Other companies crunch data from electronic monitors to evaluate employee performance, and create up-to-the-minute rankings for their bosses. Athletic companies now insert GPS locators in their running shoes to track where and how long their customers jog. Goodyear embeds sensors in tires to transmit performance data to their engineers.

These companies aren't "attention merchants" in the same way that Google and Facebook are. And they don't have entire business models built on selling and monetizing data. But they do leverage data to increase their return on investment. It's telling that the fastest way to become one of those top 10 percent of companies holding 80 percent of corporate wealth is to figure out how to leverage not physical assets or even capital, but the value of "intangible" assets, including data, patents, intellectual property, and networks. Companies in every industry are counting on such electronic data to spur growth over the next several years. Data-driven artificial intelligence could generate up to almost $6 trillion in revenues for companies that deploy it successfully. (The biggest gains now come in sales and supply-chain management.)[40] Most of the CEOs I've spoken to are extremely bullish on the subject, claiming their AI investments yield between 10 and 30 percent returns. But the more data the AI has to work with, the better it goes. That's good for corporations, but will cause a tremendous amount of disruption for citizens whose privacy is being compromised and workers whose jobs are being automated.

How is it that Big Tech has, in a matter of just twenty years, so reshaped our economy? Key to understanding that is this: Many platform technology firms operate as natural monopolies—that is, companies that can dominate a market by sheer force of their networks. Many people would argue that Google, Facebook, Amazon, and perhaps even Netflix and Apple fit this category (though Apple itself would counter that there are many competitors in its mobile marketplace, most notably Google, which takes a much larger share of the overall mobile market if tallied by percentage of users on the Android system). Natural monopolies are often a product of network effects, meaning that the more users a platform has, the more attractive it is to new users. Barriers to entry, be they capital costs or simply getting there first and controlling the physical or virtual territory, are huge, and prevent others from entering the market in an effective way. That's how the railroads and the telegraph and telephone companies of the past, and even some of the media giants of today, achieved domination. For such networked businesses, monopoly tends to be less the exception than the rule, unless there is government intervention of some sort to stop it (like the government intervention with railroads and telecoms, or the Microsoft antitrust case of twenty years ago, which allowed Google to rise).[41]

As sweeping as it is, the transformation I've described has only just begun. Theoretically, each of the largest Big Tech companies operates in a separate market. But in the Darwinian struggle for market share, they so dominate their spaces that they don't just lay claim to *a* market, but seize *the* market entirely. Then, they use that power to move into new ones, creating vast meta-networks that are astonishing in their power and reach. Netflix, Amazon, and, even to a certain extent, Apple, who are relative newcomers to the entertainment business, are no longer content being the uncontested leaders in the video streaming market; now they are also dominant content producers, becoming in effect TV and movie studios, spending billions of dollars (in the case of Netflix and Amazon) on original television programming,[42] a move that has left the previous titans of the entertainment business scrambling to match them (hence the recent massive industry mergers of AT&T and Time Warner). Google has lurched into the transportation business with its bid to create a self-driving car, and Facebook is trying to launch its own finance system with the creation of a bespoke cryptocurrency, Libra (Apple has already teamed up with Goldman Sachs on a credit card).

Big Tech, in other words, doesn't just want to become a leader in one sector. It wants to become the platform for everything, the operating system for your life. This is arguably something that Amazon has done best so far. Today Amazon is so much more than "the everything store," as journalist Brad Stone called it in his book of that name. It's also a giant server farm, housing an incalculable volume of cloud storage, and a delivery service to end all delivery services. Literally. It's moved on from shipping its own products (books, socks, appliances) to just about anything else imaginable, from Netflix DVDs, Comcast cable boxes, and Condé Nast magazines, essentially taking on FedEx, the United Parcel Service, and the United States Postal Service in its ambition to become the nation's go-to for the shipping of packages and mail.

By taking over other distribution channels, it aims to be the platform for virtually *all* commerce. In the process, Amazon can cherry-pick the high end of the package-delivery business for itself, take a cut of everything else, and leave the costly, low-end deliveries to rural America to the U.S. mail.[43] Already, Amazon has commandeered over a third of the cloud's global capacity, to keep track of all of its vast operations. It even delivers unclassified intelligence reports for the CIA.

Most recently, Amazon has gotten into healthcare—a $3.5 trillion industry—working to disrupt how we buy prescription drugs, pick and purchase health insurance plans, and more, by drawing on its supply chain and trove of personal background data that could easily be supplemented with real-time reports from health monitors in homes, hospitals, and doctors' offices.[44] It is ambitions like this that have made Amazon possibly the deadliest of the killer apps in terms of sheer market power. No wonder that Jeff Bezos, with a net worth of $112 billion, has emerged the richest of the tech oligarchs—indeed, perhaps the richest person of all time.[45]

The network effect is one way the big get bigger. Another is simply by intimidating smaller players and stealing their intellectual property. I think often of a story that one Boston-based venture capitalist and serial entrepreneur told me about how when one brand-name Big Tech firm was considering hiring his firm for a data analytics venture, they asked him to create open-source code, ostensibly as an audition of sorts, and then poached his idea and took it in-house. He would speak only off-the-record, like most people in the industry, for fear of becoming persona non grata in the market.

"I had emails showing that they had taken the code," he said. "There was no way I could afford to fight them legally, but I went to my contact there and said, 'Hey, what are you thinking?' And he said, 'You have to understand, we're dealing with six times as much data per second as a large bank, but we make 1/100,000th as much profit on it. If we have to pay for anything, we don't have a business model.'"

As the tech companies get bigger and then bigger still, they are increasingly using that market power to squash their rivals, by buying up competitors as fast as possible or by poaching their talent. There's an entire sector of venture capital now devoted to funding start-ups as "talent farms" for Big Tech, rather than successful entities in their own right. Google, Apple, and others have been known to sign cartel-like "no-poach" employee agreements with rival firms,[46] effectively restricting workers from changing employers to secure better jobs elsewhere.

As awful as all this is for individual start-ups and workers, it is proving no less destructive for the economy that depends on them. For the past century, a new wave of start-ups has risen every twenty years, refreshing the ranks of the leading American firms and improving the country's global competitive standing. Not anymore. As Big Tech has risen, early stage venture capital and the number of start-ups they fund have plummeted—taking the job creation that our economy depends on right along with them. According to the Kauffman Foundation, the number of companies less than one year old declined by a shocking 44 percent between 1978 and 2012, the exact period that modern Silicon Valley was rising.[47] A number of other academic reports show the same trend line, and not just in one industry, but in all of them.[48] As the economist Robert Litan of the Brookings Institution put it in his study looking at the entry and exit of new firms into the market, "Business dynamism and entrepreneurship are experiencing a troubling secular decline in the United States." His research shows that while it's been declining for decades now, it took a particularly sharp plunge in the mid-2000s, which is when Big Tech really boomed.[49]

While there are many reasons for the trend—from demographics to mobility to immigration—many economists feel that the rise of a technologically driven superstar economy, in which a few large players have taken an increasing share of the economic pie over that time, is a big part of the story. According to the Roosevelt Institute, "Markets are now more concentrated and less competitive than at any point since the Gilded Age."[50] And, despite Silicon Valley's reputation for cranking out the New New Thing, nothing truly transformative has come out of the biggest technology firms in a decade or so; even Apple, a brand synonymous with innovation, hasn't released a new groundbreaking product since the iPad in 2010, opting instead to simply add new bells and whistles to existing product lines.[51] So where are the new innovators of today? All too often, strangled in their cribs.

GIVEN ALL THIS, one might ask why the Big Tech firms haven't yet been treated as monopolies by the federal government and broken up like Bell Telephone of yesteryear, and Standard Oil before that. Or at least transformed and constrained by the threat of regulation, like Microsoft was twenty years ago. Because of a forty-year shift in our economic thinking around antitrust policy.[52] Or, more concisely, because of one man: Robert Bork. While infamous for being voted down by the Senate in his bid for a seat on the Supreme Court (and for firing Archibald Cox in the Saturday Night Massacre of the Watergate scandal years before that), Bork has achieved far more lasting importance for his work as the author of the 1978 book *The Antitrust Paradox,* which provided the legal rationale for Big Tech's unimpeded global dominance and became the basis of a 1979 Supreme Court decision that is still being upheld today. Monopoly, Bork argued, should no longer be defined as it always had been under the Sherman Act, as a company that took unfair advantage of a commanding market position to stifle competition. Instead, monopoly occurred when a company unfairly boosted prices it charged consumers. If a dominant player didn't raise prices, according to Bork, it was not engaged in monopoly.

Big Tech firms, however, have no need to raise prices, because they have a business model by which they are not paid in money; they are paid in data, via a system of barter. And in this system, many of the rules of capitalism itself seem not to apply. Adam Smith, the father of modern capitalism, believed that you needed transparency, equal access to information, and a shared moral framework for markets to work. In the digital age, those three things are rarely, if ever, in force.

In addition to "free" or cheap products, the monopolies of today are often praised for the convenience (one of the perceived benefits of monopolies) they offer. But people tend to overlook the fact that at the same time, they also narrow consumer choice, and, more important, reduce economic competition. Data has no monetary value in the sense that the owner can't himself sell it to anyone directly (at least not yet). It isn't listed as an asset on the balance sheets of companies that grow rich from it (though many regulators believe it should be). But in aggregate it is plenty valuable to the Big Tech firms that resell it to advertisers at a staggering profit.

Exactly how much a piece of individual data is worth varies. Google and other Big Tech firms have hired most of the top data economists in the world, which means that there's little neutral or transparent research being done to reveal just how valuable that data really is. But one recent study, done by the security analysis group Sonecon and commissioned by the Democratic strategy group Future Majority, has attempted to put a rough estimate on the wealth generated by the mining of personal data.[53] They found it was worth a whopping $76 billion in yearly revenue, not just for the usual Big Tech suspects, but for the other entities—credit bureaus and healthcare and financial firms—that mine it. The study found that sales derived from data harvesting have grown by 44.9 percent over the past two years. That's faster than in the online publishing, data processing, and information services industry itself, according to U.S. Bureau of Economic Analysis data. If the current trends hold, our data will be worth $197.7 billion by 2022—more than the total value of American agricultural output. That is resource extraction on a massive scale. If data is the new oil, then the United States is the Saudi Arabia of the digital era. The leading Internet platform companies are the new Aramco and ExxonMobil.[54]

Data is the new fuel for growth in multiple industries, from manufacturing to retail to financial services. But unlike other assets, it doesn't necessarily fuel job growth, but rather, profit growth. And those profits tend to be diverted directly into executives' and share-holders' wallets. A 2018 J.P. Morgan study found that most of the money brought back to the United States from overseas bank accounts following the Trump tax cuts went directly into stock buybacks that enrich the wealthiest people and companies.[55] The top ten U.S. tech companies alone spent more than $169 billion purchasing their own stock in 2018, and the industry as a whole spent some $387 billion.[56]

While Big Tech has done the bulk of those buybacks, and has created vastly more wealth than any other set of companies in history, they've also created many fewer jobs relative to their market capitalization than any previous generation of business giants. In 2009, the twenty most valuable companies in America had 1,790 employees per $1 billion in market cap; today they have 656.[57] Perhaps the starkest example of this trend in recent memory: When the social media firm WhatsApp was sold to Facebook in 2014, it had a market cap of $19 billion—more than any number of Fortune 500 firms—and only thirty-five employees.[58] Facebook has about a third of the employees that Google does, and Google has many fewer than Apple, which in turn creates fewer jobs than Microsoft, which creates fewer than GM. And that's not taking into account the jobs these companies disrupt—by March 2019, for example, U.S. retailers had announced more than forty-one thousand job cuts, more than double the number from the previous year, in large part due to the Amazon effect.[59]

The bottom line is that most technology businesses simply don't require many employees (think of all the robots roaming around Amazon warehouses), and this will only become truer with time. It's been estimated that globally, 60 percent of all occupations will, in the next few years, be substantially redefined because of new disruptive technologies.[60]

It's not only low-level or menial jobs that will be automated—it's all jobs. In fact, there's a case to be made that "knowledge work"—radiology, law, sales, and finance—will actually be automated *faster* than more physical jobs in areas like healthcare and manufacturing. Moreover, even in fields where humans can't be replaced entirely, the gig economy and the "sharing" economy—driven, of course, by tech firms—have dramatically increased the number of contingency workers without benefits.[61]

Beyond these relatively easy-to-track numbers is perhaps a deeper and more worrisome issue, which is the way in which data-driven capitalism has turned people into the factory inputs of the digital age. Companies used to rely on people not only as labor, but as customers that supported demand for their products (and thus demand for new labor). In the age of Big Tech, advertisers and businesses that purchase the data analytics and eyeballs are the customers. People are the product. In this sense, Google and "big data" represent a core break with the capitalism of the past.[62]

It's a shift that is almost metaphysical, as we move from an economy based on the tangible to one based on the intangible. But it is one that was perhaps inevitable, as we evolve into a new era of data-driven hyper-capitalism. Decades ago, in his book *The Great Transformation,* historian Karl Polanyi identified three "fictions" that needed to be sustained in order for the market economies of the industrial revolution to thrive.[63] First was that human life could be rebranded as labor. Second was that nature could be rebranded as real estate. Third was that free exchanges of goods and services could be rebranded as money.

In 2015, academic and tech scholar Shoshana Zuboff posited a fourth fiction for the age of Big Tech—that reality itself was undergoing the same kind of metamorphosis. "Data about the behaviors of bodies, minds, and things take their place in a universal real-time dynamic index of smart objects within an infinite global domain of wired things. This new phenomenon produces the possibility of modifying the behaviors of persons and things for profit and control."[64] Today, we live in that world, governed by our Big Tech overlords.

Feeding Our Addiction: The Cognitive Power of Big Tech

One of the reasons that we haven't yet figured out ways to curb the power of Big Tech—despite all the evidence of how it's tearing at the fabric of our society—is simply that we are too busy being distracted by the bright and shiny products and services they make. It's a cruel irony: We're all too addicted to our gadgets and apps and Facebook pages to address the problems of technology. That gets to the most invasive part of Big Tech's power: the power to manipulate our thoughts, actions, and even our brains. My son knows all about that one, but to be fair, so do most of us. According to one 2016 study, we touch our cellphones around 2,617 times a day.[65] Seventy-nine percent of smartphone owners check their device within fifteen minutes of waking up. One-third of Americans say they'd rather give up sex than lose their cellphone.[66]

I remember one Christmas Eve a few years back when I dropped my company-issued cellphone into a puddle of ice water and broke it. I tried calling corporate IT, but they had already decamped for the break. No new phone until January 2. What followed was an uncomfortably itchy detox from the 24/7 distraction of digital data. On the subway, I would find myself absentmindedly digging through my pocket for my phone. Five-minute waits in the grocery line with nothing to scroll through, click on, respond to, or "like" seemed interminable. I tried mini-meditations to distract myself during my commute. But despite all the deep breathing and visualization of stones dropping in water, my mind would quickly wander to how many emails were piling up. Melancholy set in. Without something in my hands and in my brain at all times, who was I?

There is no question about it: Our devices and the things we do on them are just as addictive as nicotine, food, drugs, or alcohol, and there is a trove of research that proves it. According to a Goldman Sachs report looking at this effect, the average user spends 50 minutes per day on Facebook, 30 minutes on Snapchat, and

21 minutes on Instagram. Add that up, and think about the effects on productivity and human relationships.

Of course, for Facebook—and the apps it hosts on its platform—this is no happy accident. It has all been carefully strategized and executed. The attention merchants want us to remain plugged in, so that they can collect more data about us and our online habits. In other words, these and many other platforms are *designed* to prolong consumption, making it seamless to go from one place of media to another—"an endless playlist on Spotify, a continuous stream of news articles on Quartz, automatic transition to the next episode on Netflix, video autoplay on Facebook . . . remove friction and consumption goes up," as the Goldman report put it. [67]

Meanwhile, sanity, of course, goes down. The American Psychological Association concluded in a recent study that "constant checkers"—those who check email, texts, and social media frequently—are more prone to stress than those who do not. Last year, a University of Pittsburgh study determined that the more that young adults use social media, the more likely they are to be depressed. I've spoken to neuroscientists who fear that the usage of mobile apps and games may lead to widespread cognitive decline or even mass early onset dementia.[68] And in China, there has been a spate of recent incidents involving obsessive online gaming, including the death of a teenager who had a stroke after gaming for forty hours straight. (The maker of the game, Tencent, subsequently instituted time limits for minors and immediately suffered a stock setback.) More troubling still, Big Tech seems to have no qualms about exploiting the mental anguish to which they contribute. Facebook, for example, has knowingly used persuasive technologies to target depressed teenagers in Australia with ads for various products and services.[69]

Given all this, it's no wonder that the backlash is starting to set in, beginning with a growing number of people who are unplugging by choice. A recent report from Ofcom, the United Kingdom's broadcast and telecoms regulator, showed that 34 percent of survey respondents had purposefully gone on a digital detox, 16 percent of respondents had purposefully gone on holiday to a destination with no Internet access, and 12 percent had chosen to leave their phones at home while they did.[70] In the United States, books on topics like "digital minimalism" and distraction-free work line the business and self-help shelves, and, interestingly, a growing wave of start-ups are working on solutions to help people resist the lure of their shiny devices. There are a number of activists and even some investors lobbying for the government to create a kind of Food and Drug Administration–style regulatory body for highly manipulative, highly addictive technology. And to be fair, some companies, including Apple and, more recently, Google, are trying to get out in front of such efforts, by making tweaks to their own devices and systems that make it easier for people to track and limit their own technology usage.

In 2018, in a speech to EU officials, Apple CEO Tim Cook admitted that there was, and is, a serious dark side to the Big Tech revolution. "We shouldn't sugar-coat the consequences. This is surveillance. And these stockpiles of personal data only serve to enrich the companies that collect them." Earlier that year, he admitted that he himself—like many Apple users—was spending way too much time on his phone, and that this was a problem. "I think it's become clear to all of us that some of us are spending too much time on our devices," said Cook at a 2018 event in San Francisco, "and we've tried to think through pretty deeply about how we can help that. Honestly, we've never wanted people to overuse our products."[71]

The outcry over what technology is doing to our brains can be heard increasingly throughout Silicon Valley, where a number of prominent tech industry insiders have finally begun calling foul. Sean Parker, the thirty-nine-year-old founding president of Facebook, recently admitted that the social network actively manipulated users' brain chemistry to keep them coming back, again and again, salivating just like 1950s psychologist B. F. Skinner's famous dogs, who were trained to expect dinner at the ring of a bell. From the beginning, he admits, "the thought process was: 'How do we consume as much of your time and conscious attention as possible?'"

To achieve this goal, Facebook's architects exploited a "vulnerability in human psychology" to create something addictive to users. Whenever someone likes or comments on a post or photograph, he said, "we . . . give you a little dopamine hit." Like most tech titans, Parker claims that he didn't really understand the implications of it all, pointing to "unintended consequences" that arise when a network grows to have more than two billion users. "It literally changes your relationship with society, with each other. It probably interferes with productivity in weird ways. God only knows what it's doing to our children's brains," he said.[72]

Tristan Harris, too, spends a lot of time worrying about the cognitive effects of today's technology, particularly on children's not-yet-fully-developed brains. Harris is a former Googler and graduate of the Stanford Persuasive Technology Lab, where he learned to engineer the kind of behavior modification software that keeps people swiping on everything from Candy Crush to potential Tinder dates to fake news. He went on to start three companies and work at Google, before going through an existential crisis about the growing power of Big Tech and founding a nonprofit that aims to curb the nefarious effects of tech addiction. As he once put it to me, "There's an entire army of engineers at all these firms working to get you to spend more time and money online. Their goals are not your goals."

The story of Big Tech is still unfurling, and each week brings as many questions as it does answers. But in my mind, the more critical question is the simplest of them all: What will we do about it?

Where Do We Go from Here?

The headwinds against solving the problems of monopoly power, addictive technology, and the political populism wrought by the largest tech firms are great. The data-driven economy is now a fact of life; companies of all stripes are counting on it to fuel their growth in the coming years. Meanwhile, Big Tech is already finding ways around whatever regulation might be coming, doing whatever it takes to continue to monetize the only product that matters—us.

It's possible that we are at a tipping point. As I write this, a number of Big Tech companies are under federal and European investigation, a topic I'll look at carefully in chapters to come. But I don't think of the tech executives as criminals. I think of them as anti-heroes whose outsized ambitions were tinged with folly, greed, and naïveté.

Much of what we resent about Big Tech should not be a surprise to anyone, least of all to its founders. The perils were all implicit in the technology we find entrancing. When Google advised its employees not to be evil, it did so because it knew full well that evil was more than a powerful temptation. Evil was baked into the business plan.

CHAPTER 2

The Valley of the Kings

There was a time when, having returned to the United States from many years as a foreign correspondent in Europe just in time for the subprime meltdown, I seriously considered going to work for Google. The 2008 financial crisis and its aftermath had been tough for many industries, but it was particularly tough for those of us in publishing and the media. Advertising had fallen off a cliff, as traditional sources of ad revenue such as the automotive industry, pharmaceutical companies, and even luxury brands tightened their marketing budgets. Newsmagazines, where I'd spent most of my career, suddenly felt like a dicey proposition. Yet I was also wary of start-ups, having had a bad experience at one in the months leading up to the first big dot-com bust. But Google, by this point, could hardly be described as a "start-up," so at the behest of a friend who'd worked for the company, I went to interview for a high-level communications job at their New York office.

The first clue that this might not be the place for me was at the front desk. As part of the automated sign-in process, I was required to agree to an NDA, or "nondisclosure agreement," in order to get a badge that would simply grant me entry to the offices. I'm a journalist, and this immediately rubbed me the wrong way. There was a lot of fine print, all of it amounting to agreeing that I couldn't speak about or write about anything I might see upstairs. I didn't even know what the job entailed, let alone what I might see—though rest assured that should I have happened to catch a glimpse of top-secret code on some careless engineer's monitor, it would have been all ones and zeros to me. Still, I decided to decline the NDA, which meant that I didn't make it farther than the company cafeteria.[1]

The famed Google canteen, it turned out, wasn't a bad place to get a feel for the corporate culture. It was filled with the predictably cheery hordes of well-groomed and affluent-looking millennials choosing from a variety of freshly made gourmet meals, all free, which were served to them by a catering staff who made sure they received no more or less than the predetermined serving size designed to minimize waste and maximize health. (You could go back for more, of course, but I was struck by the fact that Google's metric-driven ethos applied even to lunch.) The spread wasn't quite as impressive as it was at the California headquarters, where you would stumble on gourmet smoothie stations and juice bars between the beach volleyball courts and the outdoor chamber music ensemble spaces, but the mountain of fresh blueberries at the fresh fruit station seemed to have been picked minutes ago, and all in all the food was better than anything I'd ever seen in a corporate canteen. There was, of course, the requisite gourmet coffee station to cap off the meal and refuel the troops, who, between the food, the massages, the complimentary dry-cleaning pickups, and the evening entertainment sessions of lectures and cocktail mixers, seemed to have no reason to leave the office other than to catch a few hours of sleep.

I've always been skeptical of such corporate perks, which seem designed to blur the boundaries between work and life in a way that always seems to benefit the corporation more than the worker. But many people, most of whom work at the Big Tech firms, would disagree with me. As I learned from the brief conversations I struck up in line at the dessert bar, to be a Googler was more than a full-time job to many who worked there—it was a calling, one with seemingly few boundaries in terms of time or job description. The people I spoke to were rightfully proud of the company's growth trajectory and power, but to me, they also had that *Circle*-like quality (to reference the dark Silicon Valley satire by novelist Dave Eggers) of having drunk the corporate Kool-Aid. The company was good. It was kind. It wanted the best for its people and society. The fact that it was a sharp-elbowed money-making machine with what was then a 92 percent market share in search engines and an increasingly aggressive Washington lobbying presence was something the folks I spoke to didn't quite seem to have landed on.[2]

Yet the very nature of the job that I was interviewing for was indicative of a culture that was already in some disarray. It turned out that the company wanted to assign a number of experienced media or PR operatives to a handful of Google C-suiters, almost like executive body men: to follow them around, collect their thinking, and then communicate it to other executives and the company at large. It was an entirely internal PR job. I wouldn't be helping disseminate Google's message to the outside world, but rather within Google itself.

It struck me as odd, and worrisome, that the top brass running the company would have such a hard time communicating with their troops—wasn't openness and lack of hierarchy part of the Silicon Valley ethos? But the more I thought about it, the more I realized that it was indicative of an organization that was struggling to make the shift, culturally, from behaving like a scrappy start-up to being what it really was—a large, sprawling corporate behemoth that was still basically run by about five people at the top. Google thought that all it needed to do in order to successfully navigate this transition was make the wishes of these people a bit clearer internally. In fact, it needed an entirely different approach to management, one that was truly inclusive and open to criticism. While Google leadership liked to think of the company as being nonhierarchical, it seemed clear to me that power was, in fact, extremely

concentrated. And even more worrisome, the Googlers were woefully lacking in both self-awareness and awareness of how they
might be perceived in the outside world.

This is typical of not just Google, but a host of Silicon Valley
companies that have become big and powerful yet still want to act
like they are small. It's an issue that is most pronounced in the tech
sector, where the most successful start-ups are the ones that grew
very quickly. One symptom of such companies is often too much
power concentrated at the top. Facebook founder, chief executive,
and chairman Mark Zuckerberg, for example, still controls 60 percent of his company's voting rights. Recent reports suggest that he
and chief operating officer Sheryl Sandberg represent a tiny funnel
through which decisions have to flow: a management structure
more characteristic of a start-up than one of the world's most profitable public companies. Elon Musk had a similar stranglehold on
power at Tesla until the U.S. Securities and Exchange Commission
forced him to relinquish the chairmanship as part of a fraud settlement. Google has the problem, too; Larry Page, Sergey Brin, and
Eric Schmidt still own the largest chunks of the company and have
tremendous influence.[3]

I turned down the job, needless to say, and made my peace with
the fact that I was, at heart, a journalist and not a corporate flack.
It was clear to me that unless the people at the top decided the culture was to change, it wouldn't.

When Heroes Rise

To really understand the unparalleled heights of market power that
Silicon Valley holds today, you have to go back to the beginning of
this passion play, one that follows the tragic arc described by Aristotle centuries ago: the rise and fall of a flawed hero, blinded by his
own hubris. The rise is clear enough, as it starts with the moment
these dynamic firms are first conceived (usually in a garage or dorm
room, as the mythology goes) and climbs to the moment when they

go public. That's often the point at which things change—it's when companies cease to be as much about innovation as they are about share price, when their ideals take a back seat to the pursuit of as much market share as they can build up. The very best FAANG through which to illustrate this arc is the one that is most ubiquitous: Google. The story of this omniscient behemoth that functions as an auxiliary brain for so many of us today is in many ways the story of the digital revolution itself: a window into the economic, political, and social forces that explain how we got here, and where we may be going. And so, let us begin at the beginning—with Larry and Sergey.

As central as Google is to our daily lives, few people understand how the company evolved from the scrappy, cheerful, and idealistic enterprise of its early years to the vast and more ethically questionable corporate entity it is today. When I first came to the Valley as a young financial reporter in 1995, it was not to cover Google, as it didn't yet exist, beyond just a vague notion rattling around in the brains of Sergey Brin and Larry Page. I was there to see David Filo, the head of Yahoo, which had recently emerged as the Valley start-up to watch. Skinny and clad in flip-flops, he had the wide-eyed look of an Alice who'd just dropped down the rabbit hole as he showed me around the company's big open-plan Wonderland. Yahoo, which Filo had founded with his friend and Stanford classmate Jerry Yang, had been born as "Jerry and David's Guide to the World Wide Web"—it was a coder's hobby, a way to procrastinate and avoid writing their PhD thesis. But by the time I met the pair, their whimsical—if ambitious—project to pull up the world's largest collection of baseball stats had evolved into a real business, and Filo seemed somewhat overwhelmed by the whole thing.[4]

Jerry Yang was the savvier businessman of the two, although as subsequent events proved, not savvy enough. You might remember, Yahoo was a household name in the 1990s. Today, it might as well not exist, as it has a mere 1.82 percent of the online search market. In 2016, after many attempts at resuscitation (most recently by former Googler Marissa Mayer), it became the first major league player from the 1990s to be sold, to Verizon. Its demise was ultimately precipitated by two things. First, the inability to decide exactly what kind of company it wanted to be: An information aggregator? A portal? A media company? A tech firm? Second, and perhaps even more important, its inability to monetize the marvelously innovative search engine technology developed by a Valley start-up called GoTo (later renamed Overture), which Yahoo acquired in 2003.

Unfortunately, Yang and Filo had acquired not just Overture, but also the lawsuit that was under way to try to protect its intellectual property, which the original creators had neglected to enforce with patents. Because GoTo's IP could be taken with impunity, it wasn't long before a relatively new player on the scene—that is, Google—swooped in. Google's founders saw a feature in GoTo that Yang and Filo had overlooked: an auction function that would allow them to cash in on the ever-growing volume of search data by serving up hyper-targeted ads to the very users from whom that data was gleaned. This technology became the precursor to the online advertising auction system that represents the foundation of Google's business model today.

In those pre-IPO days of the early nineties, Brin and Page were spending most of their time in a cramped office at Stanford, where they were both grad students, amid computer screens, stacks of research papers, and their own budding entrepreneurial ambitions—just a pair of unshaven computer nerds building what would ultimately become the world's largest platform technology company. Brin was the cocky, outgoing one; Page the more introverted. Neither of them was getting much attention back then. In those innocent days, the coolest kid in the Valley was probably Kim Polese, the whiz kid behind Sun Microsystems' Java, a customizable, sound-and-graphics-enabled programming platform that brought the Internet to life. Kim was dubbed the "Madonna of Silicon Valley," named one of *Time* magazine's twenty-five most influential Americans on the Internet, and even appeared on the cover of *Fortune*.

Silicon Valley was a far more playful place back then, before the big money hit, and I was psyched to attend Polese's launch party for her new company, Marimba. I remember that it was full of interesting, open, energetic, and unpretentious people who seemed genuinely excited about creating the New New Thing. It was mostly guys starting the businesses back then (as it is now); Polese was one of the first women to break into the boys' club of Silicon Valley, an early prototype, as it were, of more recognizable names such as Marissa Mayer, then the Yahoo CEO, and more recently Facebook's Sheryl Sandberg, the queen to Zuckerberg's king. They were just as hard driving as the men, perhaps even more so. Mayer famously went back to work two weeks after having her first child. She and her husband, venture capitalist Zack Bogue, once confided to a writer for *Vogue* that they didn't set any boundaries between work and life. The writer marveled at how their parallel texting and emailing continued throughout breakfast, dinner, social events, and even during their interview with her.

Sandberg is the archetypal Harvard grad—stellar student, McKinsey alum, star networker, and by all accounts a tireless self-promoter. As Roger McNamee, the venture capitalist who eventually hired Sandberg to be COO of Facebook, wrote in his own book, *Zucked,*[5] "Sheryl Sandberg is brilliant, ambitious, and supremely well organized. She manages every detail of her life, paying particular attention to her image. Until 2018, she had a consigliere, Elliot Schrage, whose title was vice president of global communications, marketing, and public policy, but whose real job appeared to be protecting Sheryl's flank, something he had done since her time at Google."[6] (Schrage has since stepped down from Facebook amid the company's PR scandals.)

I was first introduced to Sandberg by a mutual acquaintance in an airport en route to Davos, where she's become a regular fixture. Her famous book *Lean In* reveals much of that sort of relentless ambition, which, I have to confess, I've always found exhausting. To me, the book seemed less an effort to address work-life balance than an attempt to brand herself as "pro-woman," perhaps in anticipation of the political career that many expect her to ultimately have, if she eventually manages to spin away her role in Facebook's privacy and election-meddling debacles.

If indeed she does run someday, it will be interesting to see how she brands herself on the campaign trail. As the election-meddling scandal has illuminated, Sandberg's core political views are, like those of so many in Silicon Valley, much more libertarian than liberal. Facebook was so desperate to protect its top leadership and its business model that Sandberg's right-hand man, Elliot Schrage, used personal clout and connections to fight off early investigations into the company's connection to Russian election manipulation[7]—even going so far as to hire a PR firm that used anti-Semitism (a particular travesty given that both Sandberg and Zuckerberg are Jewish) as a political weapon.

Indeed, it was Schrage who was on the front lines defending Sandberg and Zuckerberg after *The New York Times* broke the story, and it was Schrage who subsequently took the fall for Sandberg herself, resigning from Facebook and making a public apology in which he—quite unconvincingly—accepted full responsibility for the whole affair. As Patrick Gaspard, president of the Open Society Foundations, founded by George Soros, wrote in a letter in late 2018 to Sandberg: "The notion that your company, at your direction," tried to "discredit people exercising their First Amendment rights to protest Facebook's role in disseminating vile propaganda, is frankly astonishing to me."[8]

Being liberal in the Valley, it seems, is more about identity and less about ideology. I always found it interesting, for example, that the *Lean In* approach to gender equality seemed to put all the onus on the woman, versus focusing on the public responsibility to provide things like, say, humane working hours or decent childcare. It's a view that's common within the corporatist wing of the Democratic Party that many in the tech community gravitate toward, just like many of their "liberal" brethren on Wall Street do. (On that note, it's worth remembering that Sandberg was a protégé of corpo-ratist Democrat and "too-big-to-fail" deregulator Larry Summers, for whom she was chief of staff in the Treasury Department.)

While the Silicon Valley crew likes to think of themselves as do-gooders, they often don't make much room for the common good. It's always seemed ironic to me that even as many tech titans com-plain about the need for public sector education reform to create a twenty-first-century workforce, they also push for tax cuts and cor-porate subsidies that starve government of its ability to pay for such reform. What's true at the macro level can be seen at the micro level. I'm not the first to point out the lack of gender or many other types of diversity in Silicon Valley. Walk around any of the sprawling Menlo Park campuses or tall San Francisco towers where many tech companies now operate and you'll see few women, people of color, or, for that matter, anyone born prior to 1980. Instead, you'll see a lot of white men under forty, many of whose lack of social skills would put them "on the spectrum." These are the engineers, and they are hailed as kings.

On the surface, this makes sense. The engineers, after all, are the ones who write the code and build the platforms and design the software and hardware upon which these companies run. The prob-lem is the engineering mind-set, which focuses solely on "how do we get more efficiently from point A to point C," without much thought about the collateral effects of bypassing point B, which might repre-sent everything from the free press to citizen privacy. The result of this solutions-minded mentality is a kind of tunnel vision and cogni-tive blindness that goes a long way in explaining the lack of diver-sity, the toxic cultures, and the embarrassing PR blunders that plague so many Big Tech companies.

Gods Among Men

There is one faction in Silicon Valley whose status exceeds even that of the engineer-kings. I am referring, of course, to the VCs. If software and code are the bones of any tech company, then capital is the lifeblood, and the venture capitalists are the giant, pumping hearts that keep the blood flowing. I'm not denying that venture capital is often a necessary ingredient for innovation, or that it hasn't enabled or supported the existence of many worthy enterprises that contribute positively to society and enhance all of our lives. But any time you have a group that is held so high on a pedestal—and swimming in so much wealth—it's inevitable that at least a portion of those individuals are going to end up developing a bit of a God complex.

Consider someone like Peter Thiel, one of the infamous "PayPal Mafia," and among the first seed investors in Facebook, who went on to launch Palantir and the prominent VC firm Founders Fund. Thiel is a Trump supporter and libertarian who is critical of government and even education: Each year, he famously offers hundreds of thousands of dollars to encourage students to drop out of college and start companies instead. One of his strange obsessions is the desire to cheat death. Thiel says he finds the general population's acceptance of the prospect of death "pathological," and, along with Amazon CEO Jeff Bezos and Google's Sergey Brin, has spent millions supporting "life extension" research dedicated to "ending aging forever."[9] This, I suppose, is only slightly more ambitious a goal than those of his PayPal partner Elon Musk, also the founder of Tesla and SpaceX, who envisions supersonic commuter travel and colonizing Mars in the not too distant future (though how he'll fund it is anyone's guess, since he keeps tanking the price of Tesla's stock with his security-law-violating tweets, whiskey-and-cannabis-induced rants, and false claims about the company's financial profile).

You could argue that all of this is simply part of the "think different" mind-set, one that is necessary for entrepreneurship and radical change. The problem is that with it often comes a strong sense of entitlement and a weak sense of responsibility for any consequence of one's actions. Uber's Travis Kalanick, who became infamous for calling his company "Boober"—a crude reference to how it helped him get dates—is a great example of how this sort of tunnel vision can manifest.[10] This wasn't just adolescent posturing or "locker room talk," either; it's just one of many examples of the toxic, misogynistic culture that eventually resulted in his resignation as CEO.

Corporate sex scandals are the canaries in the coal mine of the business world: omens that foretell larger troubles plaguing the organizational culture. It hardly seems coincidental that Facebook has had its own host of sexual harassment scandals, as have Google and Amazon. When Roy Price, the top executive at Amazon Studios, was accused of sexual harassment, CEO Jeff Bezos turned a blind eye. Bezos himself was embroiled in a sordid sexting scandal in which he sent multiple penis selfies to a Fox TV personality, an episode that coincided with the end of his marriage and resulted in an ugly, high-profile legal battle with the publication that broke the story, the *National Enquirer*, which Bezos claimed had tried to extort him.[11]

My experience in business reporting tells me that incidents like these, more often than not, signal something amiss in a company's culture—particularly when they come in multiples. That was my first thought when reading in late 2018 about how Google paid the founder of its Android mobile system, Andy Rubin, a $90 million bonus as he was leaving the company, while attempting to keep quiet about one of the reasons he was leaving—a sexual misconduct claim. The details of it all had an ick factor that landed the story on the front page of *The New York Times*.[12] But it was a line in a leaked internal email response to the article, from Google chief executive Sundar Pichai to the Google staff, that really got my attention: "In the last two years, 48 people have been terminated for sexual harassment, including 13 who were senior managers and above. None of these individuals received an exit package."

Well, at least they weren't rewarded with tens of millions of dollars for their behavior; that's a positive, I guess. But really—

forty-eight people? What does this say about the company? A toxic corporate culture, to be sure. But to me, this information, in the context of the numerous sexual indiscretions of technologists at many other platform companies, was a sign of something larger: a toxic business model. That is to say, one that incentivizes these companies to tolerate plenty of egregious behavior by their top talent—assuming they are boosting the bottom line—until they are fully exposed and forced into action by the outrage of the general public.

I'm sure the majority of Silicon Valley CEOs don't condone sexual harassment, but most do seem to be rather oblivious to how they are perceived in the wider public—perhaps because they don't have to spend much time outside the greater Palo Alto bubble. Consider Elon Musk's take on riding the New York subway: "It's a pain in the ass. . . . There's like a bunch of random strangers, one of who might be a serial killer."[13]

The iconoclastic attitudes are sometimes baked in early. Marissa Mayer (who once dated Larry Page) once pointed out that if you want to understand Page and his cofounder, you had to know they both went to Montessori schools, where the philosophy emphasizes firing students' imaginations rather than just stuffing their heads with book learning. Mayer believes their unconventional educations fostered in both Googlers a willful independence and determination to go their own way, regardless of the expectations of others. As she put it to tech journalist Steven Levy in his wonderfully reported book about Google, *In the Plex,* one of the best sources for early history on the company, "In Montessori school you go paint because you have something to express or you just want to do it that afternoon, not because the teacher said so. This is really baked into how Larry and Sergey approach problems."[14]

Just how much their early educations shaped them is impossible to tell, but there's no question that their college years only reinforced this freewheeling "rules are made to be broken" ideal. Brin whipped through his undergraduate comp sci degree at the University of

Michigan in three years, landing him in Stanford's computer science doctoral program at nineteen, the youngest student ever to join the department. When Page showed up to get his own comp sci PhD in the fall of 1995 at nineteen, Brin was two years ahead of him. Later, Page said that he found Brin "pretty obnoxious,"[15] which may be his way of saying that he found Brin so impressive, he had to knock him down a peg. Stanford reeked of the competitiveness that comes from unbridled ambition masquerading as social conscience. It was the place to go if you were determined to change the world—and get rich for doing so.

Page and Brin were both involved in the Human-Computer Interaction group that eventually yielded the Persuasive Technology Lab (which I'll talk more about in chapter 6), whose work centered around taking advantage of the vast new realm of cyberspace that was just starting to be generally known as the World Wide Web. Many Stanford students saw this new virtual territory much in the same way that the first explorers to reach California must have seen the land that eventually became Silicon Valley: as highly lucrative real estate upon which things must immediately be built. Most were working to erect "portals," a point of access to news, as well as a hub from which to send email or post pictures.

Brin and Page took a radically different tack. With so much new content coming onto the Web every day—all the articles, photographs, and songs people were posting on the seemingly infinite number of new websites cropping up—they were focused on developing a way to quickly sort through it all. They understood that when the ingenious British engineer Tim Berners-Lee invented the Web back in 1989, his genius was the ability to see that all the things living in cyberspace were connected to other things. It was a welter, sure, but a welter that could be organized like any other. To him, the Web was like the Library of Congress, with each book bearing a catalog number. Only it was better than a library, as most of the documents were strung together via "hyperlinks," providing a vast network of interconnections, a web that physical libraries lacked. This Web was an unimaginably vast new frontier that could be claimed by whoever organized it first.

Brin and Page were determined to be the ones to plant that flag. At the time, "search" on the World Wide Web was a bit like trying to find a needle in a very large haystack, except that the needle was not a needle at all, but a straw of hay amid billions of other straws of hay. Fortunately, thanks to Berners-Lee, each straw on the Web bore a unique address, or URL, and most of them contained hyperlinks that connected one bit of straw to another. Still, the World Wide Web consisted of billions of items, with more pouring in every second. How could they possibly manage to organize it in such a way that would allow people to find that one specific straw they needed?

Let's say you wanted information on Tim Berners-Lee. The reigning approach of AltaVista, then the leading search engine, assumed that the document you'd most want would be the one with the most mentions of Tim Berners-Lee. Page and Brin thought that was silly. Just because the words appeared many times didn't mean it would necessarily offer the best, most useful information on the subject. But what would? Here, Larry Page relied on an insight from his parents' background in academia, where the most desirable papers on a topic were never the ones that just repeated a term or name endlessly, but the one that *other papers* cited most frequently. On the Web, the equivalent to those citations were the hyperlinks, which meant that their search engine would need a way to tally up all these hyperlinks. So Page and Brin developed a program they called Back-Rub, because it tracked links back to other documents.

Essentially, BackRub unleashed millions of tiny electronic messengers called bots to crawl all over as many documents as they could reach and tag each one with a code that only BackRub could detect and then tally up all its "back links." The resulting summary was called PageRank, an opportune pun on Page's name that was fully intended.

When they first unleashed BackRub, it burned through all the bandwidth on their departmental computers, so Page and Brin commandeered the entire Stanford University system, which had nearly five times as much. Now their bots could roam with impunity all over cyberspace, tagging, tallying—and potentially trespassing over the copyrights of anyone and everyone who had created the content they were linking to in the process, something that Google would eventually do at industrial scale when it purchased YouTube years later. (It's something they continue to try to defend with vociferous lobbying against the tougher copyright rules being pushed by both the European Union and some politicians in the United States.)

To Page and Brin, there was nothing nefarious about this. They simply sought to capture the knowledge tucked away in computer archives across the country to benefit humanity. If it benefited them, too, so much the better. It was the first instance of what later might be classified as lawful theft. If anyone complained, Page expressed mystification. Why would anyone be bothered by an activity of theirs that was so obviously benign? They didn't see the need to ask permission; they'd just do it. "Larry and Sergey believe that if you try to get everyone on board it will prevent things from happening," said Terry Winograd, a professor of computer science at Stanford and Page's former thesis adviser, in an article in 2008. "If you just do it, others will come around to realize they were attached to old ways that were not as good. . . . No one has proven them wrong—yet."[16]

This became the Google way. As Jonathan Taplin wrote in his book, *Move Fast and Break Things,* when Google released the first version of Gmail, Page refused to allow engineers to include a delete button "because Google's ability to profile you by preserving your correspondence was more important than your ability to eliminate embarrassing parts of your past." Likewise, customers were never asked if Google Street View cameras could take pictures of their front yards and match them to addresses in order to sell more ads. They adhered strictly to the maxim that says it's better to ask for forgiveness than to beg for permission—though in truth they weren't really doing either.

It's an attitude of entitlement that still exists today, even after all the events of the past few years. In 2018, while attending a major economics conference, I was stuck in a cab with a Google data scientist, who expressed envy at the amount of surveillance that Chinese companies are allowed to conduct on citizens, and the vast amount of data it produces. She seemed genuinely outraged about the fact that the university where she was conducting AI research had apparently allowed her to put just a handful of data-recording sensors around campus to collect information that could then be used in her research. "And it took me five years to get them!" she told me, indignantly.

Such incredulity is widespread among Valley denizens, who tend to believe that their priorities should override the privacy, civil liberties, and security of others. They simply can't imagine that anyone would question their motives, given that they know best. Big Tech should be free to disrupt government, politics, civic society, and law, if those things should prove to be inconvenient. This is the logic held by the band of tech titans who would like to see the Valley secede not just from America, but from California itself, since, according to them, the other regions aren't pulling their economic weight.

The kings (and handful of queens) of Silicon Valley see themselves as prophets of sorts, given that tech is, after all, the future. The problem is that creators of the future often feel they have little to learn from the past. As lauded venture capitalist Bill Janeway once put it to me, "Zuck and many of the rest [of the tech titans] have an amazing naïveté about context. They really believe that because they are inventing the new economy, they can't really learn anything from the old one. The result is that you get these cultural and political frictions that are offsetting many of the benefits of the technology itself."

Frank Pasquale, a University of Maryland law professor and noted Big Tech critic whose book *The Black Box Society* is a must-read for those who want to understand the effects of technology on politics and the economy, provided a telling example of this attitude. "I once had a conversation with a Silicon Valley consultant about search neutrality [the idea that search engine titans should not be able to favor their own content], and he said, 'We can't code for that.' I said this was a legal matter, not a technical one. But he just repeated, with a touch of condescension: 'Yes, but we can't code for it, so it can't be done.' " The message was that the debate would be held on the technologist's terms, or not at all.[17]

A lot of people—including many of our elected leaders in Washington—have bought into that argument. Perhaps that's why, from the beginning, the rules have favored the industry over the consumers they supposedly serve. The most notable example of "special" rules that benefit Big Tech is the get-out-of-jail-free card provided by section 230 of the Communications and Decency Act of 1996 (CDA), which exempts tech firms from liability for nearly all kinds of illegal content or actions perpetrated by their users (there are a few small exceptions for copyright violations and certain federal crimes).

In the early days of the commercial Internet, back in the mid-1990s, one of the refrains we heard over and over from Silicon Valley was the notion that the Internet was like the town square—a passive and neutral conduit for thoughts and activities—and that because the online platforms were, by this definition, public spaces, the companies who ran them were not responsible for what happened there. The idea was that the scrappy entrepreneurs starting message boards, chat rooms, or nascent search engines out of their basements or garages simply did not have the resources or manpower to monitor the actions of users, and that requiring them to do so would stymie the development of the Internet.

Times have, of course, changed. Today, Facebook, Google, and other companies absolutely *can*—and do—monitor nearly everything we do online. And yet, they want to play both sides of the fence when it comes to taking responsibility for the hate speech, Russian-funded political ads, and fake news that proliferate on their platforms. Apparently, they have no difficulty tracking every purchase we make, every ad we click on, and every news article we read, but to weed out articles from sketchy conspiracy websites, block anti-Semitic comments, or spot nefarious Russian bots still proves too onerous a task. That's because doing so requires real human beings earning real wages using real judgment—and that's something that platform companies that have grown on the back of automation have tried to avoid.

There have been periods when the tech giants have become more vigorous about policing for PR reasons—consider the variety of actions taken by Facebook, Google, GoDaddy, and PayPal to block or ban pornography, or to limit right-wing hate groups' use of their platforms in the wake of racially charged violence in Charlottesville, Virginia. You can argue that this is laudable or not, depending on your relative concern about hate speech versus free speech. But there's a key business issue that has been missed in all the hoopla: These companies are incentivized to err on the side of allowing content, if it will get eyeballs. They also have the power to censor. Matthew Prince, the chief executive of Cloudflare, a Web-infrastructure company that dropped the right-wing Daily Stormer website as a client back in 2017 under massive public pressure and against the firm's own stated policies, summarizes the issue well: "I woke up in a bad mood and decided someone shouldn't be allowed on the Internet," said Prince. "No one should have that power."[18]

But Big Tech does. It has *exactly* that power. It's a schizophrenia that reflects ambivalence, both on the part of the companies and society itself, about what they are. Media players? News organizations? Platform technology firms? Retailers? Logisticians?

Whatever they are, the current rules by which they play—which is to say, not very many rules at all—aren't working. The rise of Google, Facebook, Amazon, and the other platform giants has seemingly placed their leaders above the expectations, the ethical standards, and even the laws that apply to ordinary citizens. To really understand the culture, we have to dig into the business models that enabled the kings of Silicon Valley to ascend so far above others.

CHAPTER 3

Advertising and Its Discontents

Back in November 2017, a year after the election of Donald Trump, Americans got a first look at the ads that Russian groups had bought on Facebook in order to sow the political discontent that may have tipped the election in Trump's favor.[1] They made for sickening viewing. Russian linked actors had created animated images of Bernie Sanders as a superman figure promoting gay rights, and pictures of Jesus wrestling with Satan along with a caption that had the Antichrist declaring that "If I win Clinton wins!" There were calls for the South to rise again emblazoned on a Confederate flag, and yellow NO INVADERS ALLOWED signs protesting a supposed onslaught of immigrants at the border.

The images were released by lawmakers who then had a chance to question not the CEOs and decision makers who'd signed off on a business model that allowed such propaganda to be monetized, but their lawyers. As per usual, the top brass at the platforms were eager to deflect and deny any wrongdoing. The companies—not just Facebook, but also Twitter and Google—all claimed that they sent their chief counsels rather than the business decision makers because they were best positioned to respond to queries. But as their congressional testimony makes clear, the attorneys were there to make sure that the CEOs didn't have to take the fall.

"I must say, I don't think you get it," said Senator Dianne Feinstein, a California Democrat, who left the hearing feeling profoundly disappointed. "I asked specific questions, and I got vague answers." Jackie Speier, a House Democrat, also from California, summed up the situation well. "America, we have a problem. We basically have the brightest minds of our tech community . . . and Russia was able to weaponize your platforms to divide us, to dupe us and to discredit democracy."

The companies have since attended many such meetings, and in some cases sent their top brass to testify before Congress. But the message hasn't really changed. The line from the C-suites at Facebook and Google has been consistent: We are very sorry, and we couldn't have imagined that any of this would ever happen. But if you interview people who've worked on targeted advertising at such companies, this is patently untrue. The leadership at YouTube, Google, Facebook, and Twitter have known for years about the risks of platforms being misused by nefarious actors to send users down rabbit holes of propaganda. They just decided that fixing this problem wasn't worth the risks to their own business model.

The Data-Industrial Complex

A few years back, Guillaume Chaslot, a former engineer for YouTube who is now at the Center for Humane Technology, a group of Silicon Valley refugees who are working to create less harmful business models for Big Tech, was part of an internal project at YouTube, the content platform owned by Google,[2] to develop algorithms that would increase the diversity and quality of content seen by users. It was an initiative that had begun in response to the "filter bubbles" that were proliferating online, in which people would end up watching the same mindless or even toxic content again and again, because algorithms that tracked them as they clicked on cat videos or white supremacist propaganda once would suggest the same type of content again and again, assuming (often correctly) that this was what would keep them coming back and watching more—thus allowing YouTube to make more money from the advertising sold against that content. But because the subtler algorithms resulted in lower "watch time" than the original ones, the project was dropped.

Chaslot was gutted; he believed that these new algorithms would not only help mitigate the fake news problem, they would also increase business over the long haul. More diverse content, he reasoned, could open up lines of revenue that would pay off over time, as opposed to sensationalized, eye-popping content that pays off in shorter—albeit more immediately profitable—bursts. But the powers that be disagreed. Their mentality, according to Chaslot, was that "watch time was an easy metric, and that if users want racist content, 'well, what can you do?'" This was a culture in which the metrics were always right. The company was simply serving users, even if that meant knowingly monetizing content that was undermining the fabric of democracy.[3]

A spokesperson at YouTube, which doesn't contradict the basic facts of Chaslot's account, told me in 2018 that the company's recommendation system has "changed substantially over time" and now includes other metrics beyond watch time, including consumer surveys and the number of shares and likes. And, as this book goes to press in the summer of 2019, YouTube is, in the wake of the FTC investigations along with numerous reports of pedophiles using the platform to find and share videos of children,[4] considering whether to shift children's content into an entirely separate app to avoid such problems.[5] But as anyone who uses the site knows, you are, at this moment, still served up more of whatever you have spent the most time with—whether that's videos of cats playing the piano or conspiracy theories. It's true that both Google and Facebook now throw more resources at unmasking suspect accounts and removing content. But, ultimately, they do not want to be censors, and are no good at it anyway, as shown by the frequent muddles over what they do and do not decide to take down.

As for the tweaking of algorithms, Google chief counsel Kent Walker (the only high-level Googler to agree to an interview for this book) puts the company's philosophy quite simply. "We built Google for users. . . . When you're a search company, every time you make a change to any algorithm, half the people go up, and half go down [meaning the producers of content being ranked by the search engine]. And the half of people that go up think, 'Well, great to see someone's recognized how great I am,' and the people who go down say, 'Wait a second, what's going on with this.'"

Walker, who told me in an interview in January 2019 that the company had made "in excess of 2,500 changes to the algorithm last year" to stop various nefarious activity, nonetheless admits that "there's always a risk of manipulation," which is why the company sticks with its simple mantra of giving users what they want, which implies a focus on the consumer rather than the society at large.

Fair enough. But the point also drives home the power, in lieu of stronger regulation, that digital platforms like Google have to amplify humanity's worst tendencies. "Citizenship in our time," Columbia academic and Big Tech critic Tim Wu has said, "is about how you spend your attention."[6] It's a truth that has, ironically, been put into sharpest focus by Silicon Valley insiders themselves. In a speech at a European privacy commissioners conference in late October 2018, Apple CEO Tim Cook decried the "data industrial complex" made up of companies (including Google and Facebook) that make the vast majority of their money by keeping people online for as long as possible in order to garner as much of their personal data as possible. "Our own information—from the everyday to the deeply personal—is being weaponized against us with military efficiency," said Cook, whose own company still makes most of its money from hardware.

Apple has its own issues—from tax offshoring to legal battles over intellectual property infringement, which we will explore later. And Cook is being somewhat hypocritical when he criticizes his competitors for "keeping people online for as long as possible," given that Apple tries, with some exceptions, to do that, too, particularly via its promotion of the "freemium" gaming that hooked my son.

But in this particular area, it's true that other companies— Facebook, Twitter, Instagram, Snapchat, and Google—have the deeper problems. That's because their core business fundamentally depends on data mining by manipulating behavior, using an odd mix of Las Vegas–style techniques and opaque algorithms to keep users hooked.[7] These companies truly are attention merchants. We as consumers perceive their services to be free, but in reality, we are paying—unwittingly—not only with our attention but our data, which they go to great lengths to capture and then monetize.[8]

What is even more alarming, however, is how vulnerable their complex and opaque digital advertising systems are to exploitation, no matter how many people they put on the problem. The very same week that Google's $90 million Andy Rubin sex scandal hit the papers, there was news of another and perhaps even more telling debacle: 125 Android apps and websites were subject to a multimillion-dollar scam. Essentially, fraudsters acquired legitimate apps—many targeted at kids, including a number of popular games, a selfie app, a flashlight app, and more—from their developers (paid for in Bitcoin) and sold them to shell companies in Cyprus, Malta, the British Virgin Islands, Croatia, Bulgaria, and elsewhere. Unbeknownst to the users, these apps had been loaded up with bots programmed to capture their every click, scroll, and swipe—then mimic that behavior to artificially boost traffic to the apps' ads and collect bigger payouts from advertisers, even as they increased the risk of compromising the data of the real human beings who were being duped.

"The revelation of this scheme shows just how deeply fraud is embedded in the digital advertising ecosystem, the vast sums being stolen from brands, and the overall failure of the industry to stop it," BuzzFeed News reported.[9] But the fact that only Android apps were targeted highlights how vulnerable the Google platform—and, in turn, everyone who uses it—is to fraud and data breaches.

Though this story failed to make the front pages (after all, when it broke in fall 2018 the media had more pressing issues on their hands), the revelation was enough to prompt Mark Warner, a senior Democratic senator from Virginia, to write to the Federal Trade Commission, calling on it to address "the prevalence of digital advertising fraud and in particular the inaction of major industry stakeholders in curbing these abuses."

It was one of many, many letters that he and other senators have written in recent years, trying to get Big Tech to change its behavior. But simply making a few half-hearted efforts on the margins— adding a few human watchdogs here and there, or reiterating their supposed commitment to quality content over propaganda—is akin to trying to treat an aggressive cancer with a multivitamin. Why? Because these problems—filter bubbles, fake news, data breaches, and fraud—are all at the center of the most malignant—and profitable—business model in the world: that of data mining and hyper-targeted advertising.[10]

The Aura of Science

The Cambridge Analytica scandal, whereby it was revealed that the Facebook platform had been exploited by foreign actors to influence the outcome of the 2016 presidential election—precipitated a huge rise in public awareness of how social media and its advertising-driven revenue model could pose a threat to liberal democracy. But the surveillance business model itself was pioneered at Google, not Facebook, and its founders were aware of both the possibilities and the perils as early as 1998, when Brin and Page were still coming up with the name for their new venture. By the time they'd settled on *google* instead of *googol* (Page was the one who simplified the spelling—or, in one version of the tale, simply mistyped), and immortalized it in a cheerful logo, the powers that be at Stanford had begun getting very curious about this mysterious project that was siphoning off so much computer power, and reminded Page and Brin that, as academics using university resources to conduct whatever research they were getting into, they should feel some obligation to publish their findings. Page and Brin disagreed. They were too busy working on perfecting their algorithms: those complex mathematical equations that crunched data into answers.

Algorithms have the aura of science—they are based on math and quantitative information, after all. And yet, they are all too human, in that they reflect the particular ideas and biases of the people who program them. Some are better than others, of course. And already, back in the late nineties, there was a sense that the ones Page and Brin had invented were very, very good—or at least very, very valuable and thus something they should keep to themselves.

"People [at Stanford] were saying, 'Why is this so secret?'" Terry Winograd, who was mentoring both Page and Brin at the time, recalls wondering at the time. "'This is an academic project. We should be able to know how it worked.'"[11]

This attitude highlights a fundamental difference between the academics and entrepreneurs. Academics are rewarded for revealing the findings of their research—ideally in a peer-reviewed journal—so others can learn from them. Entrepreneurs, on the other hand, need to keep proprietary secrets that could yield big money. It was becoming abundantly clear that Page and Brin were decidedly the latter.

Page, in particular, was wary of being scooped, citing the cautionary tale of Nikola Tesla, the brilliant Serbian scientist who in 1931 had been celebrated on the cover of *Time* magazine for his innovations in robotics, electricity, and radio, but who had died in poverty because he failed to commercialize his ideas. (Tesla is largely remembered today because Elon Musk paid the otherwise forgotten engineer tribute by naming his electric car after him.) Page vowed that he would not go the way of Tesla.[12]

Still, Winograd prevailed, and in 1998, while still at Stanford, Page and Brin published an academic paper entitled "The Anatomy of a Large-Scale Hypertextual Web Search Engine."[13] Mostly, it explained the inner workings of their search engine. But it also foretold an existential conflict over how search could actually make money. The issue centered around data mining—which was Sergey Brin's area of expertise—and, more specifically, the fact that advertising driven by user data would turn out to be the way people would get very, very rich through this new invention.

Ironically, this was something that the Google founders were adamantly opposed to in the beginning. Not the data mining itself, but its marriage with search and targeted advertising. Data mining simply involved analyzing large amounts of data to discover trends and patterns in the aggregate.[14] But the idea of tracking people's individual behavior—what they searched for, which result they clicked on, and so on—and then building a database about who those people were so that the information could then be sold to the appropriate advertiser seemed anathema.

"If you read Larry and Sergey's original paper that they wrote at Stanford, where they talked about creating a search engine, they specifically said that advertising would inherently corrupt the search engine if you sold advertising. So they were opposed to the notion of having advertising on Google," recalled Douglas Edwards, one of the firm's earliest software engineers.[15]

Indeed, this view is laid out in black and white. "Currently, the predominant business model for commercial search engines is advertising," Page and Brin wrote on page 18, section 8, appendix A, titled "Advertising and Mixed Motives." But, they added, "The goals of the advertising business model do not always correspond to providing quality search to users."

In the appendix, they go on to say, "We expect that advertising funded search engines will be inherently biased towards the advertisers and away from the needs of consumers. Since it is very difficult even for experts to evaluate search engines, search engine bias is particularly insidious." This was an interesting statement, given that Google has subsequently declared that everything they do, including some of the things that have caused the greatest controversy, is for the *benefit* of users. Interesting, too, how this statement underscores the inherent complexity of the technology itself—complexity that would give Googlers plenty of room to obfuscate later on when they were asked hard questions about the very bias of which Page and Brin were clearly aware.

Of course, the future implications of this were not fully apparent back in 1998. Still, it's important to understand that even then, Page and Brin were somewhat worried. They had concluded that the risks of malfeasance in commercial search were not insignificant. They even considered whether search should be left in the public domain, where it wouldn't be as easily manipulated as it might be under an ad-based business model. But the pair ultimately concluded that the downsides of private sector search were "likely to be tolerated by the market," or, in other words, *people either wouldn't know or wouldn't care that they were being manipulated.*

And for a while, they didn't.

A Million Clicks a Day

By the time Brin and Page published their paper, their friends and colleagues at Stanford were banging out more than ten thousand search queries a day. Though their site was still relatively unknown outside of a small circle of academics and tech geeks, its traffic was growing exponentially. Stanford wasn't going to kick them out of the lab, since there was a certain pride associated with the new invention, but Page and Brin needed money to take their venture to the next level. Luckily, they had the Stanford network right at their fingertips.

They approached a professor, David Cheriton, who connected them with Andy Bechtolsheim, a Stanford alum who had gotten rich cofounding Sun Microsystems.[16] Bechtolsheim saw immediately that the fastest way to make money was via advertising, and in particular the targeted advertising that would appear alongside search results. Bechtolsheim recalls thinking, "Well, we'll have these sponsored links and when you click on a link, we'll collect five cents. And so I made this quick calculation in the back of my head: 'O.K., they are going to get a million clicks a day at 5 cents, that's $50,000 a day—well, at least they won't go broke.'"[17] Then he handed Brin and Page a check for $100,000, got back into his Porsche, and drove away.

He was the first. Subsequent early stage investors would include some Silicon Valley legends, including Amazon founder Jeff Bezos, who put in $250,000 in 1998. "There was no business plan," said Bezos to journalist Ken Auletta in a *New Yorker* article. "I just fell in love with Larry and Sergey."[18]

And so Google, the company, was born. That first year, they embodied the caricature of a young Silicon Valley start-up, complete with pilgrimages to Burning Man, fancy perks to lure top talent, and plenty of hanky-panky. "Sergey was the Google playboy," remembers Charlie Ayers, the first company chef, whom Page and Brin hired when the company had only twelve employees.[19] "He was known for getting his fingers caught in the cookie jar with employees that worked for the company, in the masseuse room. He got around. HR told me that Sergey's response to it was, 'Why not? They're my employees,'" Ayers recalls in Adam Fisher's book, *Valley of Genius*.[20]

In the meantime, big things were happening. The biggest of all was that the huge amount of search data that Google was generating had begun to create a self-reinforcing cycle, whereby users begat users, because in the end, search wasn't so much about the brilliance of a particular algorithm, but the amount of data it had to work with.[21] In other words, Google seemed to have captured the holy grail for any fledgling Internet company: the network effect, which simply means that the more people use a product or function, the better it becomes. The mechanism was almost ridiculously simple—a reputation for the best search would draw more users toward their site, and more users meant more data—which would build not only a better search engine, but also eventually revenue.

Google had up until that point been earning most of its money from licensing its search technology to various content sites, slowly but surely accumulating the traffic that would beget more traffic. Moreover, all of this user behavior would now be tracked via "cookies," digital tags that saw where people went and what they did online. The cookies didn't ID people by name or address, but they did provide information for a database that would prove extraordinarily useful for the company's nascent advertising business. After all, the more you knew about what people were doing, and when, and where, on the Internet, the more you could make selling that information to advertisers who wanted to reach them exactly where they were.

Even as Google was gaining traction, there were still plenty of bigger search engines with more users out there, most notably Excite, which had been funded by yet another Sun Microsystems alum, venture capitalist Vinod Khosla. And of course, there was Yahoo. The latter was arguably the shiniest dot-com start-up in the Valley at that point, the Swiss Army Knife of websites that dabbled in everything from search to email to content aggregation and creation. Yahoo decided that rather than focus on the nuts and bolts of search itself, it would focus on simply being a "portal" to the Web, and outsource search to Google.

That was Google's big break, which came in June 2000. In addition to the millions of dollars Yahoo paid them, the deal granted Google a major concession—users of Yahoo's new search function would see Google's logo and a message that Google was, in fact, powering the search.[22] But the most valuable part of the deal for Google wasn't the money or the brand recognition—it was the data. Not only would it provide the raw material that made the search engine itself smarter, and thus help it continue to attract new users, it would also fuel what would soon become the equivalent of a money-printing machine—the online advertising system known as AdWords.

Selling advertising, as we've learned, had not been part of the original plan (to the extent that there was an original plan). But while the millions from the Yahoo deal were nice, Google needed more than just a onetime hit of cash to make its investors really happy; they needed a consistent revenue stream. What good were all these users, anyway, if you couldn't make money from them? The company was burning through $25 million a year, and its VCs were getting anxious about the business proposition.[23] "There was a lot of pressure to generate revenue," recalled early Googler Douglas Edwards, and so Page and Brin decided that maybe they'd been too restrictive in their moral philosophy. Maybe it wouldn't be so bad after all to put paid advertising at the top of the search results. The two decided "advertising doesn't *have* to be evil—if it's actually useful and relevant," said Edwards.[24]

The idea for what eventually became AdWords was the brainchild of a man who was already becoming a Silicon Valley legend for his Midas touch: the serial entrepreneur, investor, and visionary Bill Gross. A brilliant Caltech grad, Gross had founded Idealab, an "incubator" that rolled out start-ups the way movie studios rolled out films.[25] (At press time, Gross had launched 150 companies and done forty-five IPOs or merger and acquisition deals.)

When Google was founded in 1998, Gross had just started a company called GoTo.com, a kind of Yellow Pages of the net, which allowed businesses to compete in an auction for the advertising space next to the most relevant Web pages. For example, advertisers of, say, mountain bikes, would bid on the search terms *mountain bikes*, and the highest bidders would get to display their ad alongside the results. When GoTo first debuted its technology at a TED conference, people were impressed but also horrified—Gross was tampering with the supposedly objective search results by allowing advertisers to pay for ad space that made them look higher up the

rankings than they were. This was so counter to the prevailing ideology at the time that people actually hissed during Gross's demon-
stration.[26] Undeterred, Gross went on to cut huge deals to provide
his service to AOL.com and other companies, netting $50 million in
revenue.

Meanwhile, the Google guys—and even more important, their
investors—were watching intently, and what they saw was dollar
signs. "GoTo prospered, and Google executives took notice,"[27]
wrote *Wired* cofounder John Battelle, author of *The Search,* which
outlines the development and commercialization of Google's search
function. Google investor Michael Moritz in particular had become
worried that the existing revenue model of licensing search technology to other companies simply wasn't going to work, calling it "a
brutal path."[28] Why try to make money deal by deal, rather than by
leveraging the power of big data and advertising?

Here again, the network effect was key; more data meant better
search results, which meant more advertisers, which meant more
traffic clicking through to more ads, which meant more data, and so
on. He pushed Google to look closely at GoTo's technology as a
model, and they did. "People started reading about how much
money was being brought into various other companies by search
advertising, and it was kind of decided that we were leaving money
on the table," noted Ray Sidney, Google employee number five.[29]

He had a point. The network effect was proving to be formida-
ble. Google had powered 3 million searches a day in August 1999—
and by the summer of 2000, that number was up to 18 million. Add
in the searches they were powering for Yahoo, and that number rose
to 60 million. But those searches and the user data they yielded
hadn't yet been fully monetized by targeted advertising—people
simply didn't click on sponsored links the way they would if the
links paid for by advertisers actually appeared to be part of the or-
ganic search results, as they did in the GoTo system. So the Google
founders "cleverly fastened on the proposition offered by GoTo," as

Google VC Moritz put it. Had they not "adopted some of the advertising techniques that were working for others, [Google] would have ended up a small, but nice, high-end company."[30]

The result, as we now know, was a company that would come to dominate the entire search and advertising industries, and set a business model standard for the platform technology companies as a whole. Targeted advertising would become the key way to make money not just for Google, but for Facebook, Amazon (which uses targeted advertising to lure users to its own site and prompt them to click on things they may have been interested in before), and a host of other companies. It created an entirely new model for how to get people to buy things.

Surveillance capitalism would become the best and fastest way for companies to grow—not just the platforms themselves, but all the other sorts of businesses that would advertise through them. That's a key point. While it would only be a matter of time before companies like Google would enter all kinds of other industries— from healthcare to transportation—the business interests that might have protested their increasing power in the marketplace remained silent, because they were benefiting from the targeted advertising model that the platforms delivered. Companies of all sorts could now spy on customers 24/7, and target them ever more precisely. It was a Faustian bargain, since they were giving up more control than they were getting, and allowing the Big Tech firms to grow fat in ways that would eventually come back to bite them, as Silicon Valley began to enter *their* markets.

But the promise of an economy that ran on data, rather than dollars, was simply too enticing to resist. The attention merchants had risen—and the race was on to capture as many eyeballs as possible, and keep them online for as long as possible. The filter bubbles, fake news, and challenges to liberal democracy that would come from that would be a problem for other people, at another time.

CHAPTER 4

Party Like It's 1999

'm no big fan of public confessions. But at this point in the book, it feels important to admit it: Reader, I used to work for an Internet company. In 1999, pretty much anyone under forty with loads of ambition was looking for a way to get into the Internet business. Internet usage, once a novelty, was rapidly becoming the norm, as connectivity improved and cable and telecom firms started laying down broadband fiber—millions of miles of which would be laid in the next decade. Between 1990 and 1997, the percentage of households in the United States that owned computers jumped from 15 to 35 percent. Talk of "surfing" the Web was becoming commonplace, and people began selling stuff on eBay and checking email via Yahoo and America Online. Meanwhile, Amazon was growing so fast that Jeff Bezos was named *Time*'s 1999 Person of the Year; that was the year that Americans' online Christmas shopping doubled, and Bezos took the lion's share of the orders.[1]

Markets were booming, but they were also creating what would become a bust. It was a time of easy money and low interest rates, not unlike the years leading up to the financial crisis of 2008, or indeed, the period since then. Back then, one of the triggers for the creation of the dot-com bubble was the Taxpayer Relief Act of 1997, which lowered the top marginal capital gains tax rate in the United

States from 28 percent to 20 percent, and in turn made more people more interested in becoming speculative investors. Fed chair Alan Greenspan had actually encouraged this by talking up stock valuations, but that would, ironically, only help facilitate what he himself called "irrational exuberance" in the market. All of it was made possible in some senses by the Telecommunications Act of 1996 and other pro–Big Tech laws that allowed Internet firms to avoid many of the pesky regulations that other companies had to deal with.

It was around this time that I got my own opportunity to jump on the dot-com bandwagon, as—no, I'm not joking—a venture capitalist myself. The fact that a bunch of millionaire investors were willing to hire a journalist who'd never worked in either technology or finance to scout "pan-European B2C media deals"—meaning "business to consumer" for those who don't remember the jargon of those days—and give her a six-figure salary and thousands of stock options to do so was clearly the sign of a market top. That seems obvious now, but was less clear to me back in the fall of 1999, when I was wooed by the London-based high-tech incubator Antfactory, a start-up venture group comprising former investment bankers and emerging market investors whose Panglossian goal was to become the Idealab of Europe. Antfactory didn't have a Bill Gross–type innovator at the helm, but had nonetheless managed, in those days of easy money and overconfidence, to somehow convince bigger fish to give them $120 million to invest in European dot-com start-ups.

They found me because I had just written a cover story for *Newsweek* on Europe's Internet boom. "Europe's Got Net Fever," screamed the headline. "The Symptoms Are Clear—Hot New Companies, Job Offers Featuring Ferraris and a Certain Swagger as Bright Young Businessfolk Stake Their Claims in Cyberspace." Yes, it's a tad bit embarrassing to write these words now. If I had thought much about it, I'd have realized that I was coming in just as the roller coaster was beginning to shift into free fall. The covers of American newsmagazines are generally a reliable counter indicator of what's happening in the markets, and this one was no exception. The market slide began a little over a year after that cover ran.[2]

I'd gotten the idea for the cover during a trip to London with my then-husband, Kambiz, who'd grown up in the United Kingdom after his parents fled Iran following the 1979 revolution. Like most expat Iranians in London, he hung out with a well-heeled group of hipsters, and through these friends, we had been introduced to some of the players in the burgeoning London tech scene. In London, the tech elite congregated in the poshest parts of the bustling city, rather than the bland, suburban enclaves you find in Silicon Valley.

It was at the peak of the "Cool Britannia" era, and the scent of euphoric optimism, thinly veiled snobbery, and, of course, money was in the air. New Labour (which was Britain's version of the corporatist wing of the Democratic Party, then powerful in the United States) still seemed like a good idea, and Tony Blair was widely regarded as the Bill Clinton of the United Kingdom (in a good way). Oligarchs looking for tax breaks and Americans looking for quick deals and easy money were flocking to the city, and everything—most notably property prices—seemed to be moving in a single direction: up. Business plans were being sketched on bar napkins in private members' clubs like Soho House and Home House, an eighteen-room Georgian mansion on the chichi Portman Square in London's Marylebone neighborhood: places where only rich expats and "toffs," the dismissive term for Britain's own native upper class, could afford to live and play.

Home House was, in fact, where I had my first interview with the partners of Antfactory, who had contacted me following my cover story, which was picked up widely in the British and international press. The partner who made the initial phone call was a South African transplant to London, Rob Hersov, a onetime Rupert Murdoch protégé who had launched a curious venture called Sportal, a pan-European sports "platform" (translation: website) that worked with local sports stations and teams like Italy's Juventus, Germany's

Bayern Munich, and France's Paris Saint-Germain. Despite the brand names, there wasn't really much there. Yet Rob himself was quite a marketing man. His most memorable features were a stunning head of wavy blond hair and a megawatt smile. He reminded me of a Ken doll, albeit one that drove a Range Rover and had a socialite wife.

Rob had rung me up in the weeks following the story and made an interesting proposition: Would I like to come and be a part of a great new London-based start-up that was looking to build "scale" and "synergy" in the as-of-yet-fragmented European Internet market? Unlike the United States, which already had a captive base of 250 million consumers who spoke the same language and wanted to buy more and more of the same stuff online, Europe was a collection of countries, each with a unique culture and individual markets. The euro had only just begun to unite the underdressed young partygoers who wordlessly undulated to the beat of pounding techno music in clubs from Ibiza to Berlin. They didn't need words—optimism was the common language.

This was the moment when Europe would finally and truly come together in an economic and political union that was to be the greatest-ever experiment in benign globalization. And Antfactory, Rob explained, would be right in the center of it, leveraging all these wonderful new economies of scale. It would find the best European dot-coms—in travel, in music, in finance, in health—and mash them together, boosting their global presence and stock valuations accordingly. Exchanges from London to Frankfurt to Paris would compete to take these new companies public. Riches would be had by all. It was so *logical*—after all, why have six different travel-booking sites (serving France, Italy, Spain, and so on) when you could have one?

From where I sat—in a dark, low-ceilinged newsmagazine office in midtown New York, where I spent hours upon hours rewriting prose sent in by correspondents from abroad—this seemed like an appealing proposition. In the States, companies like Yahoo, eBay, and Amazon had begun their meteoric ascents, and the migration from print publications to online ones was well under way—spawning glossy new magazines and sleek new websites dedicated to the digital economy, like *Wired, Fast Company,* the *Industry Standard,* and *Red Herring.* At *Newsweek,* we were all beginning to suspect that our world wouldn't be the same for much longer. True, free dinner was still being served to editorial staff in the wood-paneled penthouse floor that had once been a conference room for General Motors. But the company had, much to the dismay of "the old-timers" who had been around long enough to experience the creature comforts of an earlier era, finally started cutting back on the expense-able (and often excessive) drinking on the late nights that we'd go to press. That, too, was the sign of a seismic market shift.

So were the fortunes of people like Laurel Touby, one of the free-lance writers for the last magazine I'd worked for, a business publication called *Working Woman.* Laurel had left journalism to start up a networking group for creative types working in online media, which she called mediabistro. Laurel wasn't a star writer. But she was an incredibly capable entrepreneur, and a gracious hostess at the salons she organized at Lower East Side bars, wearing a brightly colored feather boa so that people could identify her. I attended some of her events, which were fun, and we would occasionally hang out and lament, over coffee or a drink, about the decline of the media industry and how we could barely afford to live in New York City on a newsmagazine salary. I remember once saying to her, "Laurel, you should really give up journalism and start a business." In fact, she did, eventually growing her informal networking group for freelancers into mediabistro.com, a go-to resource for writers, reporters, and editors that cornered the market for media and PR job listings and was later sold to a tech firm for a cool $23 million. Laurel, who owned 60 percent of the company, promptly bought an apartment so stunning that it was featured in *The New York Times,* got married, and became a start-up investor herself.[3]

The idea of trading in the punishing demands (and even more punishing paycheck) of the reporting life to partake of the riches of the Internet was certainly swirling around my own mind as I considered Rob's offer. I already had plans to relocate to the United Kingdom—my husband's father was sick, and the idea was for us to move closer to his parents' home in Brighton. Not only was my employer willing to relocate me, I also had offers from two competitors—*The Wall Street Journal,* and *Time* magazine's European edition—and planned to make a decision within the next few days. I loved my work as a journalist, and I was grateful to have options. But these jobs offered predictable trajectories—write more features, maybe become a senior editor or a columnist, watch my salary inch up by a paltry 2 percent every year. Antfactory, on the other hand, offered something new and inherently appealing to a business journalist—the chance to get into the game, rather than just write about it. I told the Antfactory guys (they were all guys, of course) that I'd come meet them in London and discuss the offer. I obsessed over which outfit to wear to the meeting, eventually deciding on a fifties-style retro suit and a pair of Sigerson Morrison T-strap heels, which was indicative of the fact that I knew a hell of a lot more about fashion at that point than I did about technology.

The partners, as it turned out, didn't know much more. Rob had worked mostly in entertainment and the luxury brands industry. Two others were former finance guys, one from Salomon Brothers and another from Morgan Stanley. There was a handsome but morose American named Charles Murphy, who wore a Savile Row suit and seemed unusually formal in both dress and demeanor for a would-be dot-com guy. And then there was the founder, Harpal Randhawa, a fast-talking Indian investor who had made his money in emerging markets in ways that I didn't quite follow. But the new CEO, another Rob named Rob Bier, made sense to me. He was a cheerful American with a can-do spirit who'd been a partner at the consulting firm Monitor. He would, like Google's Eric Schmidt, be the "adult supervision" at the Antfactory.

I was a good enough reporter to see that none of them were top-shelf. Still, the idea of being able to waltz into a high-tech incubator as a partner and scout deals at the height of the dot-com boom—and in London, by any objective measure a better place to live than a Silicon Valley suburb—seemed like something worth thinking about. I figured there was probably a 60 percent chance that the operation would eventually go belly-up. But the partners had enough money to fund it for a couple of years, and I had no children or serious financial commitments at the time. Plus, the stock options they were giving me might actually be worth something. And in a worst-case scenario, I told myself, I would go back to business journalism knowing a whole lot more about the tech industry than I did before I worked in it.

Ferraris and a Certain Swagger

We moved to London in December 1999, and I immediately plunged into the dot-com scene, where I soon encountered the London equivalent of Laurel's brainchild—a networking salon–cum–headhunting site called First Tuesday. It had been founded by Nick Denton, a former *Financial Times* journalist who'd eventually go on to start Gawker, and a Silicon Valley transplant named Julie Meyer. Both of them were more businesspeople than technologists. The real engineering on this side of the pond was happening in Cambridge, England, or in Estonia (where there was a thriving community of cheap coders), and Scandinavia (which was already big into mobile), whereas the DNA of the London start-up scene reflected the DNA of the city itself: It was about money and dealmaking.

Of that, there was plenty. New jobs—and new companies—were being posted on First Tuesday's website every few minutes. Within a few months, the corner bar where Nick and Julie held their monthly mixers could no longer accommodate the hundreds of eager minglers who would regularly spill out the doors. By 1999, the operation was expanding to twenty-five other cities. And there were investors lining up to give it money. The idea that venture capitalists would pay millions for a cut of a venture that served solely as a gateway to other ventures seems like an "only in America" kind of story. But the entrepreneurs and investors who made up First Tuesday felt themselves to be, as Nick Denton put it to me at the time, "spiritual Americans," people who blended American charisma with European style.[4]

At the First Tuesday networking events, Denton and Meyer made it easy to see where the money was—potential investors wore red dots on their lapels, and the "talent" wore green. The reds, who were rarer and inevitably more popular, would often emerge exhausted after a couple of hours of fighting off eager greens over cheap wine and snacks provided by the organizers. The group claimed that $100 million worth of deals had been cut in this manner. I didn't really believe them, but I didn't entirely *not* believe them. Either way, one thing was clear: There was money to be made. And it was always a fun scene, full of glittery people.

Some of the city's better known success stories included Ernst Malmsten, who looked a bit like a Nordic Elvis Costello, and his partner Kajsa Leander, a former Elite model. Back in Sweden, they'd started bokus.com, at one point the world's third largest online bookseller. After selling the company in 1998, they'd turned their attention to the fashion market, specifically targeting the young, tech-savvy customer who they thought would pay full price for tough-to-find sportswear items like Vans sneakers or Cosmic Girl T-shirts. Turned out that Bernard Arnault, the head of the global luxury conglomerate LVMH, thought so, too, as did the Italian retailer Benetton. Malmsten and Leander's Boo.com quickly raised three large rounds of capital, opened offices on Carnaby Street, and launched simultaneously in seven countries.[5] Another good-looking and posh pair, Old Etonian Brent Hoberman and Martha Lane Fox, the daughter of an Oxford historian, started Lastminute.com, a successful travel site that specialized in great deals on eleventh-hour vacations. They eventually took the company public in a £577 million listing.[6] Even First Tuesday itself was eventually sold for roughly $50 million.[7]

Meyer, to her great credit, admitted that First Tuesday had been "in the right place, at the right time." That was rare in the dot-com world. Although I had met many smart and savvy entrepreneurs who didn't make it for one reason or another, those who *had* made it all had something in common: They invariably chalked their success up entirely to their own inherent abilities. The idea of luck was all too often dismissed as a factor.

Many of the entrepreneurs bopping around London in those days were savvy Americans, often Silicon Valley types looking for less-trodden pastures where fortunes could still be made. Some, like Flutter.com cofounder Josh Hannah, an affable Stanford Business School graduate who played in my husband's poker group, had come to set up sports betting sites, something you couldn't do so easily in the United States. Josh had come over a few years earlier, with just a business plan and his hard-driving American work ethic. His shop was like all dot-coms and many tech firms today—filled with nerds in T-shirts and flip-flops, seemingly carefree and even slightly goofy, but actually as rapacious and laser focused as any high-powered corporate types. I remember getting in fights with him about the need for work-life balance, which he eschewed and claimed (perhaps correctly) wasn't really possible in a start-up. All the hustling paid off for Josh—Flutter ended up merging with the online bookmaker Betfair.com,[8] and the company later went public in a $2.2 billion offering on the London Stock Exchange.

He wasn't alone. By the end of 1999, the stories coming out of London were starting to rival the tales from the Valley, even if the companies themselves did not. I remember hearing about a hiring battle for a particular executive who was being courted by the London dot-coms. The funders of one financial services website invited him to lunch. When he got there, his would-be employer asked if he noticed all the cars in the parking lot. Then he pointed to one—a Ferrari—and told the exec that if he signed with them, he could have the keys on the spot. The executive replied that he didn't like red, and went and signed with another eager start-up.

It was the roaring '90s after all, and not just in Silicon Valley. If the 1980s began the age of greed, then the late 1990s cemented it. It was the age of Clinton and Blair, who continued much of the market deregulation that had started under Reagan and Thatcher. It was a period when the last vestiges of the labor movement and the old-fashioned notion of retiring comfortably with a gold watch and a pension began to slip away, replaced by the glamorization of Gordon Gekko and soccer moms reading *Money* magazine and hoping to become stock-picking millionaires overnight. Wall Street guys were raking in money, but not nearly as much as the Silicon Valley geeks who had the trappings of money, and yet also somehow a greater sense of having earned it, by creating real value in the form of their companies. Yet most of that value would prove to be on paper only.

Sand Hill Road venture capitalists were pouring millions into Silicon Valley's dot-com companies; British investors tried to emulate them. Continental banks desperate for higher margins were also trying to get in on the easy money. European players like the Dutch bank ING, and Credit Suisse First Boston, a subsidiary of the Swiss financial giant, were aggressively going after new deals, even bringing in investment bankers from the United States to help them identify the hottest targets.

Antfactory was trying to do the same. But it was already becoming clear that the people in charge didn't really have the skill set to pull it off. Most of the opportunities we were looking at were copycats of successful start-ups that had already launched, or ill-advised attempts to create Internet arms of existing legacy brands. I was asked to shepherd a project called Peoplenews, for example, which was basically an online version of existing glossy gossip magazines. But the whole thing seemed pointless; we weren't helping to create anything truly innovative, just trying to gin up something with "dot-com" at the end of the name. I remember coming into work day after day and feeling that people were just pretending to be busy, researching fruitless ideas in the open-plan spaces that I will forever believe are actually counterproductive to getting real work done (who can think with people constantly talking around them?). Although the founders kept a deal flow going and the press kept writing naïvely positive stories about London's homegrown tech incubator,[9] internally, the firm was already starting to revert to what it really was—a collection of ex–City bankers looking to make a quick buck.

When you pull back the lens, that's really what much of the late '90s/early 2000s dot-com boom was all about. Far away in Silicon Valley, there were a handful of firms, like Google, Amazon, PayPal, and others that were carving out sustainable niches—and then capturing that market. And then there was everyone else. As Google's Hal Varian put it in an "Economic Scene" piece for *The New York Times* in 2001, "The obvious corollary to winner take all is loser gets nothing, and there will inevitably be many more losers than winners."[10] True enough. But the frenzy of financial activity was being fueled by more than just your typical market forces. In retrospect, it reflected the increasingly tight links between the world's financial capitals and technology hubs, and the halls of power in places like Washington and Brussels. And as the economist Mancur Olson had so presciently warned decades earlier, it would be the beginning of big trouble for the political economy as a whole, in the sense that the monied elites were well on their way to buying out the political system.[11]

While the tech industry wouldn't majorly ramp up its lobbying efforts until the mid-2000s, Wall Street and Washington had a few years earlier successfully lobbied to pass new rules concerning stock options that would support the inflated valuations being thrown around like dice at a craps table, thus helping to inflate the bubble to epic proportions. It was a shift that came under the Clinton administration, which received a tremendous amount of support from both Wall Street and Silicon Valley. Bill Clinton, still one of the best politicians of all time, had managed to bring together a broad coalition of support from both the progressive end of the party, who liked his campaign message about bridging the inequality gap, and the pro-business camp, who liked his free-trade, laissez-faire approach. His team included proponents of both—Joseph E. Stiglitz, the progressive economist and Nobel laureate, headed up his Council of Economic Advisers, whereas neoliberals Bob Rubin and his deputy Larry Summers snagged roles at the Treasury. (Summers's own deputy chief of staff was Harvard grad Sheryl Sandberg, who, as we've already seen, would leverage the "market knows best" thinking to great effect later on at both Google and Facebook.)

Those camps would eventually come into conflict over stock options—the paper money that had become the lifeblood of Silicon Valley and legal tender in the casino that the dot-com boom would end up being. More specifically, it was a debate that would center around the contentious issue of stock buybacks (when corporations bid up the price of their own shares by buying them back on the open market), which had been considered illegal market manipulation until the Reagan administration legalized it in 1982. But the practice didn't really become a key part of a dysfunctional system of skyrocketing corporate pay and bad corporate decision making until the 1990s, when "new economy" tech firms began successfully lobbying the Clinton administration against efforts to introduce new accounting standards that would have forced them to mark down the value of stock options on their books.

In other words, firms wanted C-suite executives to be able "to buy company stock at below-market prices—and then pretend that nothing of value had changed hands," as Stiglitz once pointed out. It's a mark of how strong the financial and tech lobbies had become that their efforts were supported by key Democrats, including California senators Barbara Boxer and Dianne Feinstein, as well as most conservatives.

The Clinton administration was supportive, too. It introduced rules that would cap tax-deductible CEO pay at $1 million, but granted an exception for "performance-based" pay over $1 million, thus opening the door to even higher bonuses delivered in the form of stock options. Stiglitz believes this was one of the more problematic legacies of Bill Clinton's tenure. "When they pushed through the tax exemption for performance pay," he says, "they made no effort to ensure that the increase in stock prices was in any way related to performance. The favorable treatment was granted whether the increase in stock prices was a result of the efforts of the manager or the result of a lowering of interest rates or a change in oil prices."[12]

Making matters worse, the tax code, which was gradually relaxed to favor corporate debt over equity (corporate margin debt is today at record highs thanks to the tax benefits of borrowing), gave companies even more incentives to manipulate their share prices with buybacks. "The whole stock options boom caused so many incentives for bad behavior of all kinds, and for making each [corporation] look better than it was. It's all directly responsible for what I'd term 'creative accounting,' which has had such a devastating effect on our economy," says Stiglitz.[13]

The buyback issue would reemerge as an even bigger problem after the financial crisis of 2008 when companies like Apple and Google would take advantage of the ultralow interest rates (which themselves were a response to the crisis) to issue loads of bonds on the U.S. debt markets and then use the proceeds to pay back the richest shareholders in the form of buybacks and dividends, thus increasing the wealth divide.[14] But back in early 2000, a different problem was emerging—the dot-com boom was turning to bust. The value of the NASDAQ index peaked on March 10, 2000; three days later news that Japan had once again entered a recession triggered a global stock sell-off, which led to the usual "flight from risk," in which investors start to dump fundamentally weak stocks whose problems had been previously masked by "creative accounting." On March 20, *Barron's* ran a cover story entitled "Burning Up: Warning—Internet Companies Are Running Out of Cash, Fast." Companies were starting to issue reversals of revenue statements, and investors began to realize that many once-lauded start-ups were more style than substance. Once the Fed decided to raise interest rates, the die was cast. The "easy money" had officially run out.

DotComDoom.Com

Everyone remembers companies like Pets.com going under, but there were thousands more companies that either folded or got acquired in the months and years following the downturn—eToys .com, Excite, Global Crossing, and iVillage, to name a few.[15] Marimba, the company founded by Kim Polese, the mistress of Java whom I'd interviewed years before during my first trip to Silicon Valley, flamed out just a couple of years after Polese was named one of *Time* magazine's "most influential Americans" in 1997.[16] And in London, the aforementioned Boo.com shuttered its offices after spending a good chunk of its $125 million in capital on advertisements in glossy magazines, in-office champagne, first-class airfare, and lavish parties—while apparently neglecting to invest in building a site that actually worked. In fact, there were so many dot-coms going belly-up that entire websites were devoted to chronicling their woes. Yahoo's home page linked to something called the dot-com Flop Tracker, and in a case of cruel irony, the U.S.-based DotCom-Doom was one of the few sites that was thriving, with growth figures in the high double digits.[17]

Those were the legitimate failures of the late '90s and early 2000s. Then there were the outright frauds that people began to pay attention to, as they always do, once the market had bottomed out. In both the United States and Europe, those years saw regulatory investigations and hundreds of lawsuits against investment banks, analysts, and technology firms, and many of Europe's most prominent firms got caught up in scandal, with accusations ranging from knowingly promoting bad stocks to insider trading to bribe taking. For example, shareholder activists sued Deutsche Bank for dumping

44 million shares of Deutsche Telekom just two days after one of its own analysts put a buy recommendation on the stock.

A number of European and Asian investors even turned to notoriously tough Manhattan attorney Melvyn Weiss to press class-action suits against Wall Street banks and tech firms, including a number based in Europe. Credit Suisse First Boston, for example, was named in 49 out of 138 of Weiss's suits, charged with having inflated commissions and taking what amounted to bribes from clients who wanted to guarantee themselves a piece of a hot IPO. Lots of big-deal players—including CSFB's controversial technology investment banking boss Frank Quattrone—went down.[18] Stars like Henry Blodget, the tech research analyst from Oppenheimer who first predicted Amazon's success, were accused of pushing tech stocks that they knew to have problems. Blodget was ultimately convicted of securities fraud, had to pay a $2 million fine and a $2 million disgorgement, and was banned from the industry. He later reinvented himself as a technology journalist.

By the end of 2000, I was wishing that I was still one, too. And this was before things even hit rock bottom; the implosions of Enron and WorldCom were still to come, and although the markets were way down, they hadn't yet hit their lows (by the end of the dot-com downturn, stock markets would have lost $5 trillion in market capitalization). But Antfactory had become a dismal place to work. Charles Murphy, who was downbeat on a normal day, had sunk into a depression after his wife, a tiny blond American social climber named Heather, left him for Sol Kerzner, an aging South African casino tycoon with a perma-tan. The debacle was all over the London tabloids, which delighted in reporting that Heather had taken their children and moved to Paradise Island, one of Kerzner's Caribbean resorts. Every morning as I walked by Murphy's glassed-in office, I would see him sitting there with his head in his hands. Even Rob Bier's good humor had evaporated, replaced by a cynical "get what you can while the getting is good" attitude. Unsurprisingly, the cadres of young consultants and bankers who'd once clamored to join the company quickly thinned out.

The party seemed to be well and truly over. I decided to leave Antfactory, and return to *Newsweek* as the European economic correspondent in what had been my plan B all along.

And not a moment too soon. By September 2001, Antfactory's investors had lost patience with the leadership, and demanded that they return a $120 million cash pile and wind down the firm. I heard, years later, that Rob Bier had gone on to become an executive coach for Asian CEOs in Singapore, while the always sharp-eyed and sharp-elbowed Harpal Randhawa had gotten into the Zimbabwean diamond business. They fared better than most. Charles Murphy, who later moved back to the United States and worked a series of high-profile hedge fund jobs, became embroiled in the Bernie Madoff scandal after it was revealed that a firm he had worked with in the years following Antfactory had invested money with Madoff on his recommendation. Unable to keep up his glamorous lifestyle, which included a luxurious Upper East Side townhouse, he sank into despair. In March 2017, he threw himself to his death from the twenty-fourth floor of the New York City Sofitel.[19]

Is This Time Different?

I often think about those disastrous years and wonder what has changed in the tech world, and what hasn't. Today's tech market is so much more developed, with vastly better infrastructure and truly game-changing innovations. We are only just beginning to move into artificial intelligence, the Internet of things, and other areas that many businesses are counting on to propel revenue growth in the future. Whether they will or not remains to be seen. But it's a fair bet that machines talking to one another will have a heck of a lot more practical and productive applications than online gossip websites did. Certainly, many of the companies created in the past decade will have more staying power than those of the preceding generation; the Internet itself has simply become the fabric of our economy in a way that creates scale and opportunity.

But that won't be true for every company. There are ways in which the state of today's tech industry is worrisomely similar to that of the late 1990s, in the sense that easy money and a glut of copycat consumer ideas have created a bubble that has already started to deflate. Consider the lackluster IPO of Uber, which I will cover later in this book, or the rise and fall of Jawbone, the nearly defunct maker of wearable technology that not so long ago was handing out its brightly colored fitness-tracking wristbands like lollipops to VIPs at Davos, only to spend the past few years selling itself off piece by piece. You could argue that the company was a victim of being too early into the market (it launched Bluetooth-enabled devices in the late 1990s), or realizing too late that functions like sleep monitoring and step tracking would soon be performed by apps that lived on iOS and Android platforms, rather than by stand-alone technologies that would warrant their own devices and ecosystems.

But you could just as easily argue that this Silicon Valley unicorn, which at its peak boasted a valuation of $3.2 billion and attracted money from the world's most successful venture capitalists, such as Sequoia Capital, Kleiner Perkins, Andreessen Horowitz, and Khosla Ventures, was a victim of its own success. In a classic case of what's been dubbed "the foie gras effect," the company burned through so much money and reached such sky-high valuations that it became, perhaps like one of its users, too rich and fat for its own good.

Jawbone had to turn to the Kuwait Investment Authority for cash just to stay afloat, never a good sign, given that sovereign wealth funds are not exactly the smart money in Silicon Valley.[20] They tend to come in big but late, offering loads of cash when others will not, or when start-ups want to inflate their valuations prior to an IPO. Indeed, many of the Big Tech platform firms that took money from the Middle East have come to regret it. Uber, for example, which received funding from the Saudi government, went to great pains to distance itself from Crown Prince Mohammed bin Salman, the autocrat accused of ordering the murder of journalist Jamal Khashoggi (a charge that he naturally denies), by awkwardly pulling out of a Saudi investment conference known as "Davos in the Desert" (along with a number of other high-profile U.S. businesspeople) right after that horror broke.

The demise of companies like Jawbone and the lack of excitement about new IPOs are just two signs of the bubble economy in the Valley. Burgeoning debt is another. Netflix, for example, recently raised $2 billion through a junk bond offering to fund new content.[21] It will be interesting to see how the next round of big anticipated IPOs goes—or if they go at all. Many top tech companies have opted to stay private longer, bidding up their valuations and raising expectations. Both Uber and Lyft completed disappointing IPOs as I was finishing this book. I suspect they won't be the only companies unable to live up to the hype. I'm thinking in particular of Elon Musk's SpaceX, but also Peter Thiel's Palantir, which has been scaling back its thirteen-course lobster tail and sashimi lunches in anticipation of its public offering (probably a good idea, given that the company has yet to turn a profit in its fourteen-year history, despite having a valuation of $20 billion).[22] Today's darlings can so easily become tomorrow's discards; as I finish this book, SoftBank, the bloated Japanese tech investment firm, has just scrapped its $16 billion plan to buy a stake in WeWork.

Valley veterans smell the froth. "In some ways, Jawbone reminds me of Palm [the former personal digital assistant maker], in the sense that there was a real market there," says investor Tim O'Reilly, the chief executive of O'Reilly Media and author of WTF? What's the Future and Why It's Up to Us, a book about the role of Silicon Valley in our bifurcated economy. "But in another way, it reflects the financialization of the tech sector. Jawbone rode a wave of enthusiasm that was ultimately speculative in nature and reflective of the 'tulip' quality of the tech market right now."[23]

It seems not so much has changed since 2000, when start-ups like Pets.com were able to go public and jack up share prices even as they were losing hundreds of millions of dollars. Yes, the digital ecosystem has since grown, changed, and deepened. And yes, today it is harder for companies to receive funding just by sticking *.com* behind their names. But now, as then, you do not necessarily need profits—or even paying customers—to draw investor interest. All you need are "users" in a hot market niche. I've always felt that there was some measure of ingenious scam in the whole paradigm.

Still, as Citibank's former CEO Chuck Prince once quipped, "As long as the music is playing, you've got to get up and dance." Over the past five or so years, there's been massive growth in the number of these venture-capital-backed "unicorns"—start-ups with a market capitalization of more than $1 billion. Low barriers to entry have resulted in many competitors and a race to spend as much as possible to grab market share. Not only do the private companies that emerge from this unproductive cycle become bloated, so, too, do the venture funds themselves. Billion-dollar venture funds, once unheard of, are now commonplace. Last year, Sequoia raised an $8 billion seed fund, and SoftBank a whopping $100 billion fund.

Big, of course, begets big. Charismatic CEOs and their PR teams drum out compelling narratives around these sectors (wearables, electric cars, the "sharing" economy, cybersecurity). Then they send market signals about their own "value" with announcements that play off these narratives: Uber's $680 million purchase of self-driving truck firm Otto, for example. Venture capitalists and private equity investors keep the bubble going by buying into it at higher and higher valuations. As more and more heavyweight VCs bid up the value of start-ups, others have to follow. It's up or out.

The result has been not only a new bubble in IPO markets, but the undercutting of a host of public companies that actually *do* have to worry about profits. Two classic examples are Uber's disruption of the taxi industry and Airbnb's of hotels. This may be good for some of the VCs who can use the inflated values of the unicorns on their books to raise more money and charge more management fees. But I can't see how it is good for economic value overall.

Meanwhile, that capital, generated by valuations that are based as much on narrative as fact, is used less for R&D or as an investment in growth than to pay nosebleed salaries. As these pay packages skyrocket, of course, so does the price of property, services, and labor. You'd weep if you saw the price tags on depressing prefab ranch-style homes off Highway 101, which runs through Silicon Valley. The whole cycle is straight-up "madness of crowds," as described by Charles Mackay in his seminal study on crowd psychology published back in 1841. The problem with this herd mentality is that typically only a few firms win, and those winners are likely to be the small number of companies that can use their network effect to capture and control data and user ecosystems. That's why I look skeptically at the valuations of most technology groups, private and public, aside from the platform giants (and even those depend on the prevailing regulations and rules of digital trade not changing).

There are many things about the current economy that remind me of my time working in venture capital in London. Then, as now, we were in the late stages of a credit cycle, with too much money chasing too little value. And then, like now, investors were counting on a spate of hot IPOs to pour a little more kerosene on markets that were clearly overinflated. We all know how that ended, on both sides of the Atlantic. That's not to say that there wasn't value created then, as there has been now. For every unsuccessful dog food retailer or expensive T-shirt purveyor that went out of business in the dot-com bust, there were miles of broadband cable laid, which created the infrastructure that Google and other companies now capitalize on. Today, the digital economy has conveniences and economies of scale where before there were none.

But the bubble today is, in important ways, bigger and more dangerous. Venture capital money collapsed after 2000, came back up, fell again after the financial crisis of 2008, then rebounded to record levels after 2014. While technology has made starting a company cheaper, becoming a success is now much, much more expensive. That is because of an arms race to build the next unicorn start-up. As University of California academics Martin Kenney and John Zysman put it in "Unicorns, Cheshire Cats, and the New Dilemmas of Entrepreneurial Finance," their paper on the shifts in start-up funding, "Start-ups are each trying to ignite the winner-take-all dynamics through rapid expansions characterized by breakneck and almost invariably money-losing growth, often with no discernible path to profitability."

As long as investors are willing to accept growth as a metric for value, the music can keep playing. But as the University of California academics note, "Unicorns are mythical beasts." In the coming years, their financial reality, as well as the sustainability of the current funding model, will be subject to some much-needed testing. Some of the new crop of hyped-up companies may eventually turn into Cheshire cats, disappearing and leaving behind only the grins of those who got out before the bubble burst.[24]

CHAPTER 5

Darkness Rises

Before he died, Steve Jobs told his biographer, Walter Isaacson, that he intended to devote his remaining time on earth to annihilating Google's Android phone system, which he believed that Eric Schmidt—a man he'd invited to sit on his board, and whom he considered to be a close friend—had wantonly copied from Apple's iPhone. "I will spend my last dying breath if I need to, and I will spend every penny of Apple's $40 billion in the bank, to right this wrong," Jobs said. "I'm going to destroy Android, because it's a stolen product."[1]

He didn't get the chance, obviously, though he certainly tried, filing one patent infringement lawsuit after another, to no avail. (The two companies eventually settled in 2014, more or less calling a truce.)[2] As of the first quarter of 2018, Google's Android mobile operating system represented a whopping 86 percent of the smartphone universe—leaving Apple's iPhone, while still doing very well in monetary terms, a distant second.[3]

Jobs's comment may have been slightly melodramatic, but it wasn't paranoid. The truth was that Eric Schmidt *had* spent time on Apple's board: many years in fact, even after joining Google in 2001 as CEO. And it seemed clear that he had indeed acquired a number of good ideas from the company; the Android system, developed during those years, was nearly identical in many ways to Apple's iOS—an unlikely coincidence that ultimately resulted in Schmidt being dismissed from Apple's board in 2009.

But it's not as though the practice of "borrowing" ideas from a competitor was exactly outside the norm in the tech world. In 2003, for example, Facebook purchased an Israeli cybersecurity start-up called Onavo to track what competitors were doing; Facebook would then copy anything that seemed like it might be profitable. It was an internal "early bird" warning system that alerted the company to start-ups that were doing well, while giving Facebook an unusually detailed look at what users were doing on those systems.[4]

In essence, Onavo represented a legal form of corporate spying, one that produced intel that Facebook used to both undermine existing competitors and cut untold numbers of new ventures off at the pass. In 2016, for example, Facebook began paying closer attention to Snapchat, the rival that had grown wildly popular among the younger set. Snapchat's most distinguishing features were impermanence (instead of being immortalized on a "wall" until the end of time, messages sent via Snapchat would automatically disappear soon after being read) and its animated filters (a user could overlay a photo of themselves with, say, cat ears and whiskers). Coincidentally (or not) it was right at this time that Facebook's own company Instagram launched a feature called Stories, with similar features. In early 2019, Facebook announced they would shut down the digital crystal ball that was Onavo after it was reported in TechCrunch that Facebook had allegedly been paying kids and teens $20 a pop in gift cards to install the spying app on their phones.[5]

ALL OF THAT has been well documented elsewhere.[6] But Android's similarity to iOS certainly highlights how far Google had strayed from whatever idealistic roots it might once have had. By the early 2000s, as the company moved toward the inevitable payday of IPO—the dream of every Silicon Valley entrepreneur or investor—

there was little remaining pretense that Google was anything other than a leviathan of a company looking to monetize everything that it could in preparation for its debut on the public markets. As for the infamous "Don't be evil" mantra? "It's bullshit," said Jobs.[7]

This ideological shift was most publicly marked by the 2001 hiring of Schmidt. Around that time, Google was still ramping up its advertising model, and it wasn't yet clear what a gold mine it would become. The investors felt that adult supervision was needed, in the form of a hard-nosed manager who could turn the company's brilliant ideas into soaring stock prices. Page and Brin had the audacity to suggest to their investors that Steve Jobs was the only candidate they would consider, but the VCs made it clear that the two were on another planet if they thought they would get someone like Jobs—already running two public companies and a legend to boot—to come on board. They needed someone smart and savvy, but low-profile enough to play backup singer to the two founders—or at least not expect to be a rock star themselves. That was when John Doerr, a famed partner at Kleiner Perkins, suggested Eric Schmidt, the CEO of the networking company Novell, who'd previously been CTO of Sun Microsystems. He was in his forties. He wore suits. He understood bottom lines. But he was also a real engineer, someone who spoke the language of the founders.[8]

Like many Silicon Valley elites, Schmidt grew up privileged, the son of a psychologist and a Johns Hopkins professor, raised in Falls Church, Virginia. In typical fashion for those of his pedigree, he was an overachiever who earned eight varsity letters in distance running and excelled in the sciences, graduating from Princeton with a degree in electrical engineering. (He later got both an MA and a PhD from the University of California, Berkeley.) Schmidt was a tech geek, having worked as a programmer at Bell Labs and Xerox PARC early in his career. But he also had the business acumen and social skills (at least, relative to the rest of the pack in the Valley) required for management, and had quickly moved up the ranks at Sun Microsystems, where he started as a software manager in 1983.

Sun had a fast-moving culture, and one that capitalized on the new "open-source" software movement, which was all about putting code into the public domain, thus allowing developers to share work and ideas so as to build an ecosystem around a company much faster. Open-source has many advantages, and plenty of economists and technologists would say it's crucial to innovation and economic growth, because it allows entrepreneurs to build on one another's ideas; it's the polar opposite of the "walled-garden" approach of the kind that, say, Apple employs. But it can also make it difficult for companies to protect their intellectual property—a point that would become salient years later, as tech giants Google and Apple began using their power to reshape the innovation ecosystem to fit their own strategic goals.

Mr. Schmidt Goes to Washington

Within a year of his arrival at Google, Schmidt had a group in place and had created a thriving business model. Now, the trick was to protect it. Google's search engine was impressive, but in order to keep monetizing the data it generated, Google would need to make sure that it remained free, easily accessible, and unencumbered by copyrights, privacy rules, or any sort of patented intellectual property that would make it tougher for Google to capture as much traffic as humanly possible. That would require a regulatory and legal strategy, and a team of (official and unofficial) lobbyists and lawyers to do Google's bidding in Washington, D.C. So Schmidt, Page, Brin, and a small group of other insiders began interviewing candidates to head up their government policy team.

Enter Peter Harter, a top lobbyist for Silicon Valley, who had previously led government policy for Netscape and helped the company bring successful antitrust suits against Microsoft. Harter, a lawyer with a specialty in intellectual property, was a tech insider, but he also knew politics—he'd been on the front lines when a previous generation of Big Tech firms had squared off, and was

well-connected both in Washington and in the Valley. Harter had
worked with Eric Schmidt while he was CTO at Sun Microsystems,
where he'd lobbied around issues including export controls on en-
cryption software, and at Novell, on antitrust issues. He'd worked
alongside the attorney Kent Walker—whom he referred to as "the
guy in charge of s—t cleanup" at Google—when he was at Netscape,
and traveled in the same circles as Google insiders like David Drum-
mond, now chief legal officer, who'd been the tech giant's first outside counsel. Since leaving Netscape, Harter had formed his own
government affairs consulting practice, where his client list was
made up of blue-chip firms such as Microsoft. I've known him to be
an unapologetic conservative, the type who will mock vegetarians
and hypocritical liberals (humorously, I must say), but also as someone willing to work for whoever could afford him. So in 2002, when
a Google executive named Omid Kordestani came calling, asking
him to weigh in on the privacy issues harming Google's model, he
was intrigued.

Harter eagerly accepted the proffered invitation to the Googleplex, a sprawling Palo Alto campus bigger and far better funded
than many East Coast colleges, to meet with Schmidt, Page, Brin,
and a number of other executives in a series of all-day meetings.
"Google was thinking of ways to accelerate revenue growth, and
get the best IPO valuation," says Harter. "You were starting to see
the advent of the smartphone, video coming online, and lots of
open-source, and peer-to-peer sharing, thanks to Napster [the music
sharing service started by Sean Parker, later president of Facebook,
that was eventually shut down because of copyright infringement]."
Harter says that Google could "easily see, looking at the litigation
over Napster, that they needed a growth map in Washington to get
ahead of any opposition."[9]

Harter understood that Google needed to devalue intellectual
property and prioritize access to user data in order to ensure its
supremacy—and by all accounts, the Googlers understood that, too.
In fact, that was one of their major competitive advantages. As

Shoshana Zuboff lays out in her book *The Age of Surveillance Capitalism,* Page, Brin, and Schmidt (along with Hal Varian) were the first in Silicon Valley to fully understand the concept of "behavioral surplus," in which "human experience is subjugated to surveillance capitalism's market mechanisms and reborn as 'behavior.'"[10] What she's saying, in simple terms, is that everything we do, say, and think—online and in many cases offline—has the potential to be monetized by platform tech firms. All human activity—the things we post, our videos, our books, our inventions—is potentially raw material to be commodified by Big Tech. "Google is to surveillance capitalism what the Ford Motor Company and General Motors were to mass-production-based managerial capitalism," she writes.[11] Nearly everything we do can be mined by the platform giants. But only if they can keep information free. That means keeping the value of personal data opaque, or ignoring copyrights on content, or—in the case of other types of intellectual property—by making it tougher to protect.

All of this corroborates what Harter has told me about his meeting in the Googleplex. He says that topics like antitrust policy, copyright, file sharing, and privacy were very much on the minds of Schmidt, Page, and Brin by that time. "One of the questions was, 'How do we avoid what happened to Napster?'" He outlined what he thought the Google strategy should be if they wanted to best protect the company's interests: Spend loads of money and lobbying power to make sure that Google wouldn't have to pay for the intellectual property and content that search was monetizing, and fight hard to keep liability exemptions in place, so that they wouldn't be responsible for things that users did on their platform. "I told them, 'Basically, you have to out-lobby the other guys, and prepare to litigate and generate support for your lawsuits in the media, the policy, and the political communities.' I remember Eric nodding and saying, 'I think that sounds right.'" Sadly (or perhaps not) for Harter, he didn't get the public policy position, which ultimately went to Andrew McLaughlin, who became director of global public policy at Google in January 2004, and later went to work for President Obama as deputy chief technology officer of the United States. (A PR representative for Schmidt and the other Googlers told me that they "don't remember" the entire meeting.)[12] Still, says Harter, "what they have rolled out since was basically the strategy we discussed on that day."

Innovators Versus Implementers

Google, Apple, Facebook, and others often position themselves in the public debate as "innovators," and that's true up to a point. As we've already seen, by the time these firms went public, most of their biggest and best innovations were behind them. From the IPO on, the game is more about implementing technologies—theirs and others'—to gain business model advantage. In the technology field, and increasingly in most fields, having access to the best intellectual property and data, and paying as little for it as possible, is everything. One of the ways in which Google and other Big Tech companies were able to gain an advantage over intellectual property was by pushing for an overhaul of the U.S. patent system, the first in thirty-some years, which reached its climax in 2011 with the passing of the America Invents Act.

To understand why this is important, you need to understand how patents have historically worked to protect innovators: Imagine that you are the founder of a small biotech company in the United States. You have spent millions of dollars and years of time developing a new diagnostic test for a blood disease. You are about to revolutionize your field. Following the passage of the AIA, it became harder for you to get a patent for your game-changing discovery, because shifts in the system meant that your invention was no longer protected due to changes in the list of what could and couldn't be patented, and the way in which innovators were allowed to defend their IP.

For example, even if patents were granted, following the AIA, the right to use them could be challenged in a non-court adjudication system, allowing other firms to quickly invalidate intellectual property. Unable to fully monetize their investment, many smaller companies, inventors, and innovators begin funneling their money and ideas to other places, like Europe and parts of Asia. Suppliers and talent begin moving there, too. While the story isn't one-sided, I have heard from many American investors, entrepreneurs, academics, lobbyists, and lawyers—including some of those who actually helped craft the AIA—who believe that the U.S. patent system has swung radically in the wrong direction. Over the past fifteen years or so it has moved, they say, from a system that was arguably overzealous in granting patents, to one in which the country's top minds can no longer monetize their research. That is, of course, a state of affairs that could have dramatic consequences for U.S. competitiveness in a world in which most economic value lives in intellectual property.

The shifts in the system that began a decade ago have come in a variety of ways.[13] Starting even before the passing of the AIA, there were a series of Supreme Court rulings, such as *eBay v. MercExchange* in 2006, *Mayo Collaborative Services v. Prometheus Laboratories* in 2012, and *Alice Corp v. CLS Bank* in 2014, that, coupled with the America Invents Act passed under President Obama in 2011, have made patents in the United States harder for companies without enormous legal and lobbying power to secure—and harder to defend. Perhaps as a result, many companies now complain of "efficient infringement" on the part of larger rivals, which simply copy or take the intellectual property they want, then settle with aggrieved parties out of court for less than the full value of the IP. Few companies will go on the record with their travails, for fear of being blackballed within the tech community.

How did we get here? Back in the early 2000s, when the dot-com bubble burst, many companies were left with nothing of value except their patents, which were then purchased by financial companies or larger tech entities that then tried to milk some cash from them. At the same time, the ecosystem of software suppliers that served the burgeoning commercial Internet and smartphone markets began to broaden. Pushing back on the ease with which patents could be obtained and defended was great for Big Tech, which, of course, has its own IP to protect, but which was also increasingly monetizing the data and IP created by others. The majority of those companies had legitimate technologies and ideas to protect. But some—so-called patent trolls—were playing a game of legal arbitrage, filing as many patents as possible in order to get larger companies to settle with them for the use of their technology.

By the time Barack Obama took office in 2009, the patent troll narrative had reached a fever pitch. It was a story line supported by many Big Tech companies[14] that individually and via lobbying bodies pushed for the America Invents Act. The law established a non-court adjudication body, the Patent Trial and Appeal Board. The idea was to save time and money with the non-court *inter partes* process, and indeed, patent claims went from taking three years and an average cost of $2 million to settle, to taking eighteen months and costing $200,000. The argument about patent trolls increasingly rang false. Yet the largest tech groups, particularly Google, lobbied hard for even more anti-patent legislation in 2013. Companies that supported that additional legislation said it would have cut out legal distortions around issues like the venue in which patent cases are heard, thereby cutting litigation costs.

But some regulators and lawmakers who had originally supported patent reform began to feel that the entire process was being used to push an anticompetitive market agenda on the part of Big Tech. "It was shocking to see calls from some in the tech industry for a second round of drastic patent legislation immediately after all we did in the AIA, and before the AIA had even gone into effect," says David Kappos, former head of the U.S. Patent and Trademark Office under Obama (who had supported the first round of legislation), who is now a lawyer with Cravath, Swaine & Moore. Kappos has, to be fair, represented Qualcomm, one of the critics of the cur-

rent system, which only just settled a three-continent, multiyear patent dispute with Apple. But both he and the legal firm have also
represented clients on the other side of the argument. "Ultimately,
the real agenda sunk in," he says. "This second round of drastic
cutbacks to the patent system was a commercial ploy designed not
to stop abuse but to cut supply chain costs by devaluing others' innovation."

The new legislation was ultimately held up in Congress. Meanwhile, Michelle Lee, Google's former head of IP, eventually took
over as head of the USPTO. In 2013, the White House put out an
alarming report on the prevalence of patent trolls and their destructive effects, blaming them for two-thirds of patent suits. Yet subsequent research done by the nonpartisan Government Accountability
Office put that number at one-fifth, and other data showed that the
number of patent defendants had been roughly flat before and after
the AIA. "The historical trend in litigation rates relative to patents
granted clearly does not support claims that litigation in the past
decades has 'exploded' above the long-term norm," wrote Bowdoin
College professor Zorina Khan in a 2013 paper entitled "Trolls and
Other Patent Inventions." What's more, she argued, a number of
legislative changes seemed to address "the ephemeral demands of
the most strident interest groups at a single point in time" and are
"inconsistent with the fundamentals of the U.S. system of intellectual property."[15]

Indeed, some would argue that the system of adjudication for
patents has become a shield for those accused of patent infringement. Most of the verdicts go against the patent holder, leading
Randall Rader, former chief judge of the U.S. Court of Appeals for
the Federal Circuit, the court in charge of patent appeals, to label it
the "death squad" for IP. Another retired federal district court judge,
Paul Michel, has become a vocal opponent of the system, arguing
that excessive invalidations and the way in which the adjudication
board has preempted court rulings are sapping both the strength of
the patent system and American innovation itself.

"The cumulative [anti-patent] effect of the Supreme Court rulings and the AIA was, together, stronger than it should have been," he says, in part because of what he and others say was lobbying on the part of large tech firms. "Patent values are plummeting, and licensing and capital investments in many technologies are sinking. The AIA has done more harm than good."

I can't tell you how many technologists and venture capitalists I've spoken to in the past several years who say that they simply won't invest in areas that Google or Facebook or Amazon or Apple are likely to play in, because of the difficulties inherent in protecting open-source technology, and/or defending patents against the big guys, who inevitably have more time and legal muscle on their side. As technologist Jaron Lanier has pointed out, the most profitable assets, like Google's own PageRank algorithms, or the closed system of the iPhone, are almost always proprietary, rather than open. "While the open approach has been able to create lovely, polished copies, it hasn't been so good at creating notable originals," says Lanier,[16] a fact that underscores the way in which Big Tech firms push open-source to the extent that it aids their ability to profit from others' innovation, but rarely let competitors anywhere near the code that powers their own key technologies.

Or, as Gary Lauder, a venture capitalist and scion of the Estée Lauder family, once put it to me, "You need a patent system that induces the right behavior, which means one in which incumbents have to pay for innovations, not copy or steal them." Lauder, a Silicon Valley–based investor who has poured more than half a billion dollars in funding into nearly one hundred companies and sixty venture capital funds in the past twenty-eight years, including GoTo/Overture (the online auction company from which Google copied crucial ideas—see chapter 2), has become an outspoken advocate for a stronger patent system. "We need to protect the larger start-up ecosystem, which is where the majority of jobs are created," he says. "It's an issue that's really crucial for our economy. Today the incumbents are copying the innovators. Next both will be copied and displaced by cheap foreign knock-offs."[17]

Information Wants to Be "Free"

Then there's the way that Big Tech commodifies a different sort of innovation—the content that artists, writers, filmmakers, and others produce as part of their living. Consider the tactics deployed to weaken copyrights, which essentially could have stopped the rise of any of the Big Tech companies in their tracks, since they create almost none of the content that they so richly monetize. In fact, Google, Facebook, and other large platform players were the chief beneficiaries of the Digital Millennium Copyright Act, signed by Bill Clinton into law in 1998, which protected online service providers from being prosecuted over copyright infringement, assuming that the provider wasn't aware of the infringement. Not surprisingly, it was supported by a number of the California-based politicians and liberals that Silicon Valley interests had begun quietly funding.

The law essentially allowed platforms to bully content creators—anybody putting things online, really—into giving up their work for free if they wanted to be searchable on the biggest platforms, while at the same time accepting the fact that the platforms would be the ones to benefit most from the content, monetarily, in exponential measure. After all, just as no small innovator with a patent could battle Big Tech, there was no way that an individual writer or musician, for example, could wage a legal battle to try to get royalties from the likes of Facebook or Google—or even to understand how much money the companies were making by linking to the content or leading advertisers to it as part of their search or social business models.[18]

As an example, consider the decade-long battle between Google and myriad authors and publishers over the Google Print project, later renamed Google Books. Scanning every single page of every single book in the world had long been an obsession of Page and Brin—it was, after all, a typically Google-sized ambition. They knew that the majority of the world's books were protected under copyright from such unauthorized copying and distribution. But the Googlers felt, in typical form, that such pesky rules didn't apply to them. Plus, they couldn't understand why anyone would think it was better for authors to make money on books than for the entire world to have free access to information. So in 2002, they simply began scanning pages, albeit covertly. As tech writer Steven Levy put it in his book, *In the Plex,* which devotes twenty pages to the book-scanning project, "The secrecy was yet another expression of the paradox of a company that sometimes embraced transparency and other times seemed to model itself on the NSA."[19] Schmidt, who had by then decided that "evil is what Sergey says is evil,"[20] was all for the project, which he declared "genius."[21]

The publishing industry disagreed. In 2005, several publishers, represented by the Association of American Publishers, filed suit against Google's "massive, wholesale and systematic copying of entire books still protected by copyright." Soon after, the Authors Guild did, too, and the suits were combined. The publishers and authors wanted, understandably, for creatives with works still protected by copyright to be able to opt in or opt out of their books being scanned. But Google chief economist Hal Varian protested that this would kill the whole project, which, of course, depended on scale. Some three-fourths of the books that Google aimed to copy were still under copyright—and Varian and the other Googlers knew that many, if not most, of the authors probably wouldn't opt in.[22]

So, they decided to settle, ultimately agreeing to a compromise in which Google would agree to show only snippets of books that were under copyright for free in exchange for becoming the exclusive seller of digital copies of out-of-print books for the publishing houses and authors that agreed to the settlement. Google, which was earning about $10 billion in yearly revenue at that point, would pay the relatively tiny sum of $125 million to establish a registry of book rights holders and pay lawyers to organize the system and the payouts. It was a complete coup for Big Tech. Brewster Kahle, the head of the nonprofit Internet Archive, which wanted to do its own book-scanning project, claimed (not incorrectly) that Google had become an information monopolist. Even Lawrence Lessig, the digital law expert who favors many of the policies that the platforms support, said that Google's deal was the equivalent of a "digital bookstore, not a digital library."[23] What he means is that even as Google was presenting the entire project as being done for the benefit of users, Google itself would ultimately benefit the most. More content meant more opportunities to sell advertising.

Why did the authors and publishers ever agree to this initial settlement? Because they didn't know any better. As with the people who took subprime mortgages from big banks, there was a huge information asymmetry in the dealings of the publishers with Big Tech, which was holding pretty much all the important inside information about just how valuable the digital monetization of searching such content could be. Google understood the new digital world that it was, in fact, creating and dominating. The publishers did not, and were desperate in the short term to stop the Big Tech behemoth from eating their lunch.

Not only did they not have the information required to understand the long-term implications of content monetization via targeted advertising, they also didn't have the time to think about it. They were on the defensive in a way that many more industries today are when dealing with Silicon Valley (witness the raft of quick mergers in, for example, the food industry after Amazon bought Whole Foods). Big Tech plays offense, while the rest of us are on defense. The problem is that, as any sports fan knows, that's not a good strategy. Like most of us who were eager to use and be a part of the shiny new thing called the World Wide Web, authors and publishers didn't fully internalize the truth—that they were becoming complicit in the idea that Silicon Valley wanted everyone to believe, which was that information should be free, and that the value created by others—be it books, music, or any other type of content—was a commodity that could be mined on the cheap.

It was really only when other big companies like Microsoft and Amazon became outraged, too, that things began to change somewhat. The Google deal shut them out of a key part of the publishing market, and they wanted back in. (This paradigm has played out in a number of subsequent cases—Google and Amazon, for example, regularly do battle to try to gain more access to each other's markets, and many of the most powerful groups that complain about monopoly power on the part of Big Tech are other corporate behemoths.)

Eventually, under pressure from some 143 groups, both non-profit and private-sector, the U.S. Department of Justice took on the issue, claiming it had granted Google too many anticompetitive rights, and that the book-scanning and -selling project was a monopoly issue. Larry Page called the legal challenge a "travesty to humanity," while Sergey Brin wrote a sanctimonious piece in *The New York Times* defending Google's efforts. At court proceedings in 2010, Google's attorney Daralyn J. Durie argued that "copyright infringement is evil to the extent that it is not compensated and that it harms the economic interests of rights holders." It was a clever argument, because it shifted attention toward the fact that Google was, after all, facilitating book sales—and away from the fact that Google itself was becoming the major beneficiary of a huge amount of copyrighted content. The judge ruled in Google's favor, on the basis that the project offered "significant public benefits." That Google would now have a monopoly on the digital sale of some thirty thousand books was deemed less important.[24]

This commodification of content and transfer of wealth from creators to platform tech firms heated up further after Google's acquisition of YouTube, which specialized in user-generated content, in 2006. It was a move that amounted to a kind of industrial policy that they were offering to the world (or at least that's how one high-level Googler once put it to me).[25] Instead of getting paid for their creativity via copyright, people could make money doing stuff on YouTube. The problem is that it takes about 2 million hits, according to *Move Fast and Break Things* author Jonathan Taplin,[26] to make around $20,000 a year or so—not exactly a substitute for a middle-class job. You are either a YouTube star, or you are at the bottom of a pyramid of free labor, which critics like Taplin would say has become a zero-sum game for everyone but the technologists themselves.[27] While it's true that the new crop of tech companies makes it easier to slough off less productive tasks—driving, shopping, and so on—they also rely on "DIY" inputs, including user-generated content and open-source software. This is essentially unpaid work being done on a mass scale.

Documents from the 2007 *Viacom International v. YouTube* lawsuit provide a window into how Google saw the creators of the content it was monetizing. Email chains back and forth from top leadership (including Schmidt, Page, and Brin) illuminate how Googlers pressured the entertainment company to keep as much content as possible outside of a paywall, making it free online, where the platform could more easily make it searchable (and thus monetizable).

Putting things online for free largely benefits the platforms, not the content creators, because it means more traffic, which means more revenue. Whatever press or publicity that a content creator can gain from the exposure is minuscule by comparison, and certainly doesn't replace the revenue gained by more traditional channels of distribution; even today, it's arguably easier to make decent money as a writer or producer working for a traditional print medium or television brand than for an online outlet, with a few exceptions. But that wouldn't be completely clear to the studios (or the publishers) until much later. In the 2007 case, the major studios eventually decided it was better to be on YouTube than not, believing that the number of eyeballs involved would pay off. The courts ruled that as long as YouTube wasn't given any "red flags" by content creators in advance of posting content, then the Digital

Millennium Copyright Act allowed it to upload clips without worrying about copyright violations.[28]

For Google, these cases amounted to tens of billions of dollars in revenue gained from the work of writers, producers, musicians, filmmakers, and ordinary people who were putting content online in increasing numbers. Free, user-generated data and content is the lifeblood of platform technology companies. All of them are built on it—every tweet, every like, every search (on Google or Amazon) is the raw material of Big Tech. That's not to say that there aren't benefits to this for users, content creators, and developers—it's just that the benefits to the platform firms themselves so greatly outweigh them. Just imagine if GM or Ford had to pay for nothing but labor—no material costs, no factory costs, no cost for anything that it took to build their products, only the salaries that they paid to their workers. Then imagine that they needed a fraction of the workers that they currently have to create their products. That's the difference between a digital business model and an industrial one.

Given all this, it's no wonder that Big Tech will do whatever it takes to protect the loopholes that enable it to get around any sorts of restrictions on the monetization of user data and content—including using its megaphone to kill the Stop Online Piracy Act (SOPA), a bill that would limit access to sites that hosted or facilitated pirated content. A day after the bill was introduced, Google put what looked like a big black box across its logo, with the words TELL CONGRESS: DON'T CENSOR THE WEB. The message let people click straight through to a blank email already addressed to their congressperson, which resulted in the crashing of congressional email servers. The bill was pulled within three days.[29]

A Tide Turn?

Google, Facebook, and others tried to use some of the same tactics in Europe in 2019, in advance of an EU decision around new copyright rules (the European Copyright Directive) that will now force platform companies to take more responsibility for removing unauthorized copyrighted content, and/or pay publishers and other content creators for its usage (albeit with a relatively small fee).[30] The directive passed the European Parliament by a wide margin, despite a smear campaign by the Big Tech firms, in which they tried to make it seem as though the new laws would penalize small companies who couldn't comply with the rules effectively, and/or impinge on freedom of speech.[31] A top German newspaper, the *Frankfurter Allgemeine Zeitung,* published an undercover exposé showing that Google money was linked to YouTube "activists" who had organized to protest the laws. It turns out that far from being part of a grassroots movement, they were being paid by Google to protest[32]—not unlike the way in which poor people in countries like, say, Iran, are paid by the government to show up and decry this or that idea that might be unfavorable to the regime.

Meanwhile, actual content creators were happy to have the European government legitimize the idea that their work should be properly protected and paid for by the people using it. "[Platforms now] have an incentive not to upload content that violates a copyright, and they have an even greater incentive to sign licensing deals with the owner of that content," Thomas Rabe, the chief executive of the German media group Bertelsmann, told the *Financial Times.* "We have to put an end to this for-free culture on the Internet."[33]

The fact that the EU has taken some steps toward protecting content creators is hopeful. But it's certainly not the end of the story. Spain, for example, tried to enact similar measures unilaterally in 2014, and Google simply "turned off its news service there," according to the editorial board of the *Financial Times.* In Germany, it has opted to "carry news only from sites that agree to have content shown for free."[34] But the success of the platforms at the expense of content creators has come at a cost. Ironically, Facebook and Google both recently launched services to try to support local news, because after years of struggling in the wake of their business model being destroyed by the platform firms, there simply aren't that many local news outlets left. That, in turn, begins to affect the platform companies themselves—if there's nobody left to create content, then what's left to monetize?

The disruption has taken a huge toll on news media and publishing, which have been most directly affected by the rise of the Big Tech industry. Newspapers have essentially been in decline since the mid-1990s, when the commercial Internet took off.[35] Today, over 62 percent of the U.S. population gets news from some form of social media, with Facebook being the dominant source, according to the *Columbia Journalism Review.*

But looking at the patent issue makes it clear that there is a price to be paid for the "information wants to be free" approach in other industries as well—including the technology business itself. While the complexity of global supply chains and investments makes it tough to show clear causality between patent protection and innovation in the United States, the trend lines do not look good. According to one study, the shifts in patent regulation have cost the U.S. economy $1 trillion. Venture capital investment in biotech has been down in recent years, in part because it's tougher to patent certain kinds of innovation. Anecdotally, I've spoken to many investors who have said they are considering moving money away from the United States and into Europe and Asia because of this. Those are, of course, highly skilled jobs that the country should be looking to keep.[36]

Cartels and Collusion

Large technology companies use cartel-like methods to protect what is perhaps their most valuable resource: their employees. Consider the agreements among a number of Big Tech firms not to poach one another's employees. When companies attempt to lower their potential labor costs by agreeing, via a backroom handshake, to make it tougher for workers to leverage one offer against another, that behavior is typically considered collusion. When it happens in Silicon

Valley, however, it typically also means that one or both companies are trying to prevent their top talent from absconding to a competitor—and taking their proprietary information, ideas, and secrets along with them.

In 2011, court documents revealed that in 2007, after Steve Jobs called Google to complain that a recruiter was trying to hire one of his people, Schmidt wrote an email to HR saying, "I believe we have a policy of no recruiting from Apple. . . . Can you get this stopped and let me know why this is happening? I will need to send a response back to Apple quickly."[37]

While at Google, Schmidt instituted a "Do Not Call" list of companies that could not be tapped for talent, something that was legally dicey, since it basically undermined the ability of individuals to look for work. And clearly, he knew it: According to one court filing, when Schmidt was asked in an email by Google's HR director about the no-poaching agreements with competitors, Schmidt responded that he preferred it be shared "verbally, since I don't want to create a paper trail over which we can be sued later."[38] As one key staffer for a senior Democratic senator later said to me, "Those guys should have gone to f—ing jail for that." Instead, Google, Apple, and two other firms implicated in the scandal—Adobe and Intel—agreed to pay $415 million in damages in an out-of-court settlement.

As Peter Harter put it when I interviewed him about his experience with the Googlers, "These companies get so powerful and have so much money and access to politicians and employee talent, that the usual rules just don't apply."[39]

IF THE LARGEST players can't steal or poach the IP they are after—they simply shell out a few million and buy it. Google, for example, has purchased more than 120 companies in the past decade. (Facebook and Amazon have purchased 79 and 89, respectively.) Such purchases, however, are just as likely to be defensive ploys as offensive ones. Facebook's 2014 acquisition of Oculus, an up-and-coming virtual reality start-up, for example, was much less about getting into that business than killing the upstart's promising operating system, which might have eventually competed with their own.[40] And its acquisitions of Snapchat, Instagram, and WhatsApp were about making sure that no one else would develop a new social network that Facebook followers might jump to. Amazon's Alexa is based on the technology of a start-up that got a mere $5.6 million from the company, who then copied the voice assistant pretty much wholesale.[41] Google, of course, is the biggest buyer of all, with more than two hundred acquisitions made during its history.[42]

Given the rapaciousness of the giants, and the fact that Washington has done little to nothing to stop such mergers (since they aren't perceived to be in conflict with the Bork-era "consumer welfare" concept of lowering prices, a topic we'll examine in chapter 9), it's no wonder that there are now innovation black holes around anything that the platform technology companies are (or might be) interested in, something known in the business as "kill zones."[43] As several venture capitalists and technology executives have told me, nobody wants to start a company in an area that they know Google or Facebook or Amazon might be entering, unless it's to create a kind of "talent farm" that might eventually be bought up by the giants simply for human capital. This isn't innovation building or creating new jobs. It's about making existing giants richer.

And richer they have become, to the tune of double-digit margins, year after year, with vastly fewer input costs than most other businesses. Apple, in less than four decades, became the world's first trillion-dollar firm, with others following closely behind. Google went public on August 19, 2004, with an $85 floor. By the end of the day it was around $100, and 900 Google millionaires had been created. By day two, it was $108. And by February 2005, it was at $210. It was all up from here for Google, and no wonder. As the IPO and the financial documents revealed during that time would show, Google wasn't just a tech company. Page 80 of the prospectus laid it out: "We began as a technology company and have evolved into a software, technology, Internet, advertising and media company, all rolled into one."[44] AdSense was a money-printing machine. Yahoo, seeing the writing on the wall, dropped its patent lawsuit in exchange for shares in Google. Big Tech had come of age. In the years to come, Big Tech would grow even bigger and more powerful than anyone could imagine, thanks to the new science of persuasive technology known as "captology."

CHAPTER 6

A Slot Machine in Your Pocket

At the beginning of this book, you read about how many of my anxieties about Big Tech came home to me—quite literally—in early 2017, when I opened my credit card bill and noticed quite a number of small charges from Apple—$1.99 here, $5.50 there, and so on. At first I figured I must have downloaded a song or a movie and forgotten about it. But then I noticed that there were three or four a day, and sometimes as many as ten or fifteen, for many of the days in the month, so many that if I'd somehow incurred all the charges without remembering any of them, I would have taken it as a sign of a neurological problem. I checked the previous month's bill and saw a flurry of small charges there, too, though there were fewer, so I hadn't noticed. I checked online to see the current bill; the most recent charge had been made that very day. I pulled out a calculator to add them all up, and I was stunned to see that the total came to a whopping $947.73.

What? My first thought, of course, was that I'd been hacked. Some enterprising computer nerd sitting in a basement somewhere (or, alternatively, a ring of nefarious cyber-criminals with ties to Moscow) had broken into my account, and was draining it in tiny increments. But as you read in the author's note, it soon occurred to me that I was not the only one with access to my Apple account. I'd given my then ten-year-old, Alex, the password—along with strict instructions that he always check with me before he used it. Perhaps he knew something about this?

I went downstairs and found him in his usual position on the living room couch, iPhone in hand.

I sat down beside him.

"Alex," I began, gently, "have you been using my Apple account?"

Engrossed in his game, he did not look up.

"Alex, look at me."

He glanced up, the screen still flashing, phone emitting its cheery noises.

"Turn that off for a second, would you?"

With a look of mild irritation, he set the phone down beside him.

"There are a lot of charges here," I began. I started reading them off from the bill.

Alex shook his head. "Not me." I had the sense that his mind was drifting over to the iPhone, still beeping beside him.

I looked at the dates and times on the bill again, and this time I noticed a pattern—all of them were made on weekday afternoons, after Alex got home from school, occasionally unattended.

"Now, Alex, are you sure you had nothing to do with this?" I pressed.

"Wait—how much were they again?" he asked.

"A few dollars, most of them." I let that sink in. "All of them from the App Store."

He suddenly went pale. "Oh," he said. "That."

Then the story tumbled out. It had all started with a Google search. Alex had been trying to find the "funnest" soccer game on the Web, and FIFA Mobile had appeared at the very top of the list (in the paid ad portion, not that Alex realized that). Alex then went to the Apple App Store to download the game for free. FIFA Mobile was great! Really fun and exciting. But he soon discovered that he would do better at the game if he had stronger players. But alas, the real stars—like his heroes Ronaldo and Messi—were not always to be had for free. They could, however, be bought with "FIFA cash"— purchased in real dollars, of course. Then he discovered something even cooler—for more money, one could obtain packs that allowed you to arm your players and your team with even more tricks and skills—something known in the industry as "loot boxes."[1]

Sure enough, with better players, winning was easy. Alex won more games, laid on more FIFA cash, and then won more still; in $1.99 by $1.99 increments, his team ascended through the ranks as if it was the real Real Madrid. FIFA Mobile recorded Alex's scores, his stats, and where he stood in the rankings—not just against local teams, but globally. It was Alex against the world, and he was on a winning streak. The trouble was, the better he did, the tougher the competition—requiring better players, with still more skills and tricks, if he wanted to stay ahead. And, if he ever tired of the game, perhaps indulging the rare urge to turn to other things, like his homework, his iPhone would flash with a reminder of a big game coming up. It was all tremendous fun, thrilling really, for a soccer-mad ten-year-old. But even he could now see that things had gotten somewhat out of control.

When he finished telling me the story, Alex hung his head and shrugged. "I couldn't help it!" he told me glumly. "I—I don't know, the game just kind of took over." He described a kind of brain fog, a trance, in which he simply lost himself. I felt bad, but I was unwilling to let him entirely off the hook. So I split the cost with him, and arranged for him to pay off his half with allowance money, extra chores, and whatever he could make on his lemonade stand.

Suffice it to say, FIFA Mobile has since been removed from his phone.

"Persuasive" Technology

As a mother, I was horrified by this whole incident. As a business journalist, I was fascinated. How, I wondered, was this game designed to be so utterly irresistible as to turn my normally well-behaved and well-adjusted son into a veritable FIFA Mobile junkie? Was it the unique talent of one brilliant game maker? Dumb luck? Or was it something else entirely?

It was indeed something else—a very big and lucrative something. A little digging revealed that the developers who worked on FIFA Mobile did not dream up the key elements that hooked Alex. The mechanics that informed their work—a field, I would soon learn, known as "captology"—were developed years before, at the Stanford Persuasive Technology Lab. Like many other products of the digital revolution, it was an endeavor born out of high ideals, but then developed into technologies that could be deployed for baser motivations. The lab was founded by Stanford University professor of psychology B. J. Fogg, who was keen to build out the simple behavioral models of B. F. Skinner to something more sweeping. Rather than just get a rat to pull a lever for a bit of food, he sought to improve the habits of human beings.

Like many technologists that I have met, Fogg comes across as friendly but somewhat odd. This is a man who is clearly brilliant, a pioneer in one of the most important fields of study in technology today. Yet he's also a bit goofy (he poses with plush toys on his website),[2] and more than a bit naïve about the potential implications of his work. He initially reached out to me after I mentioned the lab in a column, because he felt maligned, his work "misunderstood."[3] The 2016 election-meddling scandal led to a growing awareness of how social media could be misused, and Fogg came under fire from some critics—and journalists—who felt he should have done more to warn people about the nefarious effects of technology-enabled persuasion. But it became clear to me during interviews that Fogg, like so many technologists, was more interested and able to discuss the bright side of innovation than to figure out ways to deal with the dark.

"I grew up in a tech house. We had a microwave in the garage that we'd tinker with. I'm sure it was giving off radiation," jokes

Fogg.[4] "But I was also very interested in language and wanted to do a degree in English and rhetoric. I had discovered Aristotle, and I thought, 'All this [power of persuasion] could come to technology!' " Fogg eventually pursued a PhD at Stanford, looking at how technology influenced people. "But I wanted to go further, to look at whether computers that, say, shared your personality or flattered you would be more persuasive. Nobody was studying that. Well, except the video game people were onto it."

Fogg gathered a group to study a few dozen of the persuasive technologies that were already in use, like a video game put out by the army to try to recruit new GIs, and another underwritten by Dole Food Company to try to get people to eat more fruit. The methods were grounded in the research of psychologist B. F. Skinner, who found that the most effective way to create behavioral change is through a system of what's known as "intermittent variable rewards." Back in the 1950s, Skinner discovered that lab rats are quite content if a food pellet is given consistently—every five pulls of a lever, say. But the rat will develop a mad craving for that same pellet if it is given *in*consistently: say, after one pull, then after seven, then after twenty-three. Not knowing exactly when that pellet will come, the lab rat will not stop yanking that lever. And as subsequent research showed, the same principle holds true for humans. This is, of course, the very mechanism behind the slot machine—and why it's more addictive than all other types of gambling,[5] and generates more annual revenue than all other games combined.

Variable rewards are likewise ubiquitous in the world of Big Tech. A smartphone itself is a slot machine that you carry around in your pocket,[6] and even though most of the "pellets" it feeds us are actually boring beyond belief, we keep pressing the lever, in the hopes of those random rewards: a gossipy email, a cute snapshot, an exciting bit of news. Most of the apps on our phones are slot machines, too, designed to hijack our attention, to redesign our wants, to alter our very natures. Fogg's motivations were altruistic, as he sought to encourage individuals to do good things like exercise more and smoke less. His classes in computer-human interaction, which essentially translated philosophical and psychological ideas about persuasion into algorithmic form, "wasn't about developing the dark side," says Fogg, who has become quite defensive about his role in the propagation of addictive technologies (particularly after having received death threats following some of the negative press coverage). "It was about highlighting and facilitating persuasive technology for socially useful purposes. I don't know if we were successful. . . ." His voice trails off.

That, of course, depends on how you define *successful*. Certainly, Fogg's lab became one of the most popular and influential at Stanford, churning out the founders of Instagram and a host of other start-ups. But the word really spread with the rise in popularity of Internet platforms like Facebook, which suddenly gave app developers the chance to employ the techniques of captology to influence millions. Fogg himself had built a start-up in 2003 and 2004, geared toward combating isolation by helping people—particularly older adults—build strong relationships with friends online. (AARP was a partner.) Then Facebook came calling.

"They said, 'We're launching a platform and we want you to create an app and work with us on this.' And I saw that they'd put all the pieces together. Unlike Myspace, they were starting from a position of credibility (because of Zuckerberg's Harvard connection). They'd allow people to contribute their talents and skills and have these things be part of Facebook and that would increase users and have a snowball effect of more people coming to the platform.

"It was a testing ground for new apps!" says Fogg, excitedly. "I remember getting into my little Acura Legend, and driving away and calling my mom and saying 'Mom, you've never heard of this thing called Facebook, but they just won the game. You should get on it, because this is where you'll be eventually.' "[7]

Fogg's own start-up failed to reach critical mass, though he did have an interesting meeting with Google's Larry Page about it.

"Terry Winograd was my thesis adviser and also had been his. So I reached out, and he got Page to come over, and he looked at what we were doing and said, 'You've got to get this out there. . . . You've got to get it out now.' I was concerned about making it perfect. But I think he was right."[8]

Later on, as an academic, Fogg taught a course called "Mass Interpersonal Persuasion." Facebook had 25 million users by that time (today they have a whopping 2.38 billion), and his class gave students the chance to develop apps that would be tested and marketed on the platform. For Facebook, of course, it was all part of the hunt for eyeballs. Just as any Facebook user would be prompted to share their email and phone contacts with the platform ("Facebook is better with friends!" went the pitch), so would users of any app— which meant the app's developer could eventually be able to extract information not only on those users, but also on any friends they "invited" to use the app.[9] This is exactly the way that the now-infamous data leak with the British firm Cambridge Analytica happened. The academic Aleksandr Kogan created a survey app that was deployed on Facebook, and he used it to collect information not only on the 250,000 people who actually took the survey—but also on the 87 million more users they knew. Cambridge Analytica then used that information to deploy ads that may have helped tip the 2016 U.S. presidential election in favor of Donald Trump. That's the network effect in action, at its most nefarious.

Fogg had naïvely hoped that the efforts of his students would be skewed in noble directions as they interacted with Facebook, but some had more immediate concerns, like making gobs of money. That 2007 "Facebook class," as it was known, launched seventy-five students into careers in Big Tech. "It was really hard getting approval to teach that class," he remembers. "Parents were like, 'We're sending you to Stanford to study Facebook?' But we told the administration, 'This is really important; we need to study this.'"

Fittingly enough, Fogg and his colleagues used Facebook to publicize the class. "There were about 100 students, which was a lot, but for the final presentation, 550 people showed up—many of them top investors, engineers, and innovators from Silicon Valley. It was standing room only. I was exhausted afterward. It took me, like, a month to recover."[10]

Some of the students came up with apps for health and wellness—there was one, for example, geared toward people who wanted to train to hike the Oregon Trail. Others developed dating apps. While Fogg was more interested in the benign stuff, the truth was that intermittent rewards could be deployed to persuade anyone to do just about anything—to click on a link, to stay on a page, to stay there longer, to buy things, to encourage others to buy things, and so on and so on. Over the course of ten weeks, his students engaged with a total of 16 million people on Facebook.[11]

One of Fogg's students was Tristan Harris, a young technologist raised by a single mother who worked as an advocate for injured workers in the Bay Area.[12] Harris was soft-spoken, contemplative, and abnormally bright, even by Stanford standards; as a kid, he dreamed of becoming a magician—or a psychologist. But he started to think differently when he got to Stanford and became entranced by computer science, particularly the potential for computer intelligence to improve the human variety. In Fogg's class, he studied the famous B. F. Skinner clicker training for dogs, and learned how the same techniques of intermittent variable rewards can inspire behavioral changes in people.

At Stanford, Harris began to learn how the design of online experience, or even the design of a website, could have a powerful effect on emotion. The initial design of LinkedIn, for example, publicly displayed each user's personal connections. Nobody wanted to seem like the loser who didn't have enough friends. So they began scrambling to invite more and more people, driving up the size of the network and the value of the offering in turn—and, of course, for free.[13] Social pressure, as he soon learned, seemed to be a highly effective technique for "hijacking"[14] our attention, as he now refers (quite accurately) to the process by which addictive technology works. With their infinite streams of continuously self-refreshing content—whether it's your Twitter feed or real-time updates to your virtual FIFA ranking, mobile games and apps are *designed* to make us believe there's a chance we could be missing something important. And that is what keeps us checking our phones constantly: around 2,617 times a day, according to one study.[15] We might think we are in control, but in fact, we are being manipulated by the attention merchants.

As we've already learned, this has a terrible effect on our mental health, increasing stress and anxiety levels and even risk of illness. Gaming apps—like FIFA Mobile, the slot machine that hooked Alex—in particular are increasingly categorized as addictive by mental health professionals. In 2018, the World Health Organization added "gaming disorder" to a new draft version of their International Classification of Diseases, off the back of a growing spate of research showing that large numbers of online gamers—of which there are up to 2.6 billion in the world, or one person in two-thirds of American households—aren't in control of their behavior. "I have patients who come in suffering from an addiction to Candy Crush Saga, and they're substantially similar to people who come in with a cocaine disorder," said Dr. Petros Levounis, the chair of the psychiatry department at Rutgers University, in a June 2018 *New York Times* article on game addiction. "Their lives are ruined, their interpersonal relationships suffer, their physical condition suffers."[16]

Unsurprisingly, children, who spend more time with social media, games, and apps than adults do, are particularly vulnerable. Megahit games like Fortnite, for example, which include as many as two hundred persuasive technology design tricks, have given rise to support groups for families dealing with game-addicted family members—sons in particular. Unfortunately, it's a losing battle. No matter how hard a parent might try to teach self-control, restraint, and responsibility, these lessons are no match for the dopamine hit kids get from playing the game. No wonder the designer of Fortnite admitted to *The Wall Street Journal* that his goal was to create a game that would engage kids for "hundreds of hours if not years." Epic, the company behind Fortnite, has already made $2 billion from its sales of virtual goods.[17]

Then there's the craving for social approval. As most of us will recall from our teenage years, this phenomenon is hardly new; what's new is the way Instagram and Snapchat and other platforms have elevated this need to the level of full-fledged addiction. Consider the average teenager, who spends 7.5 hours a day playing with screens and phones.[18] Is it any wonder they are more isolated, less social, and more prone to depression than previous generations?[19] As scary as this is, it's even scarier that these conditions can actually be monetized by the platforms that create them. In 2017, Facebook documents[20] leaked to *The Australian* showed that executives had actually boasted to advertisers that by monitoring posts, interactions, and photos in real time, they are able to track when teens feel "insecure," "worthless," "stressed," "useless," and "a failure," and can micro-target ads down to those vulnerable "moments when young people need a confidence boost." Think about that for a minute. It's an endless, wanton commodification of our attention, with little or no concern for the repercussions for individuals.

This is how today's devices create desires we didn't even know we had, at least not to this degree, making us feel anxiously incomplete without them, almost as if we were missing a limb. When I once asked a pal of Alex's to put his phone away for Alex's birthday party, the boy got so upset he nearly hit me. It could have been worse, as with one teen, treated by psychologist Richard Freed, the author of *Wired Child,* who became so violent after being deprived of her device that her parents had to have her strapped to a gurney and sent to a psychiatric ward. The girl's parents said her downward spiral had begun with a phone obsession, followed by isolation, falling grades, depression, violence, and finally the threat of suicide. Sadly, this is something that Freed says he is seeing more and more in his practice.

Parents don't understand, he says, "that their children's and

teens' destructive obsession with technology is the predictable consequence of a virtually unrecognized merger between the tech industry and psychology. This alliance pairs the consumer tech industry's
immense wealth with the most sophisticated psychological research,
making it possible to develop social media, video games, and phones
with drug-like powers to seduce young users." Psychologists and
social anthropologists are being hired in droves by the largest companies (and many smaller ones, too), to help the technologists trans-
late the latest persuasive research and techniques into ever more
tricked-out products designed to capture ever more children's attention.

In early 2019, a number of consumers' and children's advocacy
groups filed a request for the U.S. Federal Trade Commission to investigate Facebook for alleged deceptive practices following revelations that the company had knowingly tried to dupe children into
spending large amounts of money in online games. (According to
the unsealed class-action documents, Facebook employees actually
referred to the kids as "whales," a casino term for high rollers.)[21]
Around the same time, brands such as Nestlé and Disney stopped
buying ads on Google-owned YouTube after pedophiles swamped
the comment sections on children's videos with obscene postings.
The corporate behavior in question is, of course, wildly different in
the two cases. But the connective tissue is that children have been
endangered by a business model based on the monetization of user
content and data.

This was, again, all part of the plan. The race to capture consumer attention is the focal point of capitalism today. Of all the
states of mind that companies and brands seek to induce, addiction
is by far the most desirable. It's not enough to get people to like a
product, even to love a product. You want to make them crave it so
much they can't live without it, just as corporations did with to-
bacco, followed by alcohol and television (sex, of course, is up there,
too, but unfortunately for marketers, sex can usually be had for
free). In fact, many of today's biggest companies are finding ways to

evoke nicotine-type cravings for whatever they are selling. And many of the methods that allow Big Tech to do just that came straight out of the Persuasive Technology Lab.

"The Devil Lives in Our Phones"

Tristan Harris was a talented student, and after he left Stanford, he launched a couple of successful small companies, including one that designed the type of pop-up ads that were by then all over the Web. In 2011, he joined Google, which had acquired one of his companies, to work on designing boxes of text that would entice people to click for further information. But as he spent more time at the company, he noticed that his fellow Googlers seemed a little "off." They were easily distracted, twitchy, overwrought, burned-out—even as they professed to be joyfully engaged in their work. It dawned on him that they had many of the characteristics of drug addicts in search of their next fix. Harris tried to initiate mindfulness programs, in an attempt to combat the effects of computers on his co-workers' attention. They weren't particularly well attended, but that should have been no surprise—after all, there was work to do, and the Googlers were too busy and distracted to waste their precious time and attention on mindfulness.

"A wealth of information means a dearth of something else," as Nobel Prize–winning economist Herbert Simon once put it. "What information consumes is rather obvious: It consumes the attention of its recipients. Hence a wealth of information creates a poverty of attention, and a need to allocate that attention efficiently among the overabundance of information sources that might consume it."

The cognitive capture wrought by Big Tech is so all-encompassing, so *distracting,* it can be hard to see clearly. Certainly it's hard to think clearly, given that technology forces everyone to move at a pace that is virtually inhumane. That's one of the things that bothers Harris the most—the fact that the population as a whole is losing the ability to focus and solve problems, particularly the complex kind that we have before us today (how to fix capitalism, how to combat climate change, how to end political polarization). Forget about having the concentration to tackle those—most of us can't stay on top of our own daily email in-box or social media correspondence. Things have gotten so bad that even when we aren't using our phones, the mere knowledge of their presence still has the ability to distract us; research shows we actually perform our work better the further our cellphones are away from us (on the desk is better than in the pocket, but not as good as in another room).[22]

This research speaks to an even deeper issue: Technology is hampering the ability of an entire generation to concentrate enough to truly learn. According to data from the National Survey of Children's Health, around 3 percent of the population had attention deficit hyperactivity disorder in the 1990s. Today that's up to around 11 percent, an alarming rise that many doctors link to the rise of digital media.[23] At Columbia University, my alma mater, professors are fretting about how incoming freshmen can't focus long enough to learn the basics of the core curriculum. They simply don't have the attention span to handle 200 to 300 pages of reading per week. As Lisa Hollibaugh, the dean of academic affairs at Columbia, has noted, professors are now "constantly thinking about how we're teaching when attention spans have changed since 50 years ago."[24]

It's quite the catch-22: If we can't focus on long-form reading and absorb the information needed to forge complex ideas and thoughts, then we certainly can't solve the big-picture problems of the day—which include how to manage our interactions with technology in a way that doesn't result in myriad nefarious side effects.

It's not that the makers of these distractive technologies aren't aware of these problems; they are, in fact, quite aware—when it comes to their own lives, at least. It's quite telling that many technologists take regular digital detox breaks. Fogg was on one himself, in Hawaii, when I spoke to him by phone; he told me, "I try to stay off Wi-Fi and keep my gadgets away from me while I sleep—it drives me crazy how many people come out here and stay on their phones the whole time!" They go to great lengths to keep their own children as far away from the digital world as possible. Waldorf schools, known for their unconventional teaching methods—which include eschewing the use of electronic media and devices in the classroom—are quite popular in the Valley, as are nannies with strict instructions to police phone usage while parents are at work, busily writing algorithms and building or marketing the very devices, apps, and platforms that keep people hooked. As one parent—a former Facebook employee—put it to a reporter at *The New York Times,* "I am convinced that the devil lives in our phones."[25]

Utter hypocrisy? Or true repentance? Maybe a little of both. But it's true that a growing number of technologists are finally waking up to the sheer force of the destruction they've unwittingly unleashed—and working to absolve the sins of their past. Witness techies-turned-activists like Tim Berners-Lee, who created the World Wide Web, and is now trying to wrest it out of Big Tech's all-too-powerful hands. Or James Williams, a Googler-turned-philosopher who left the Valley for Oxford to research the ethics of persuasive technology. Or Jaron Lanier, the pioneer of virtual reality, whose recent book, *Ten Arguments for Deleting Your Social Media Accounts Right Now,* argues that social media is creating a culture of victims and diminishing diversity of thought in a way that will undermine not only our economy and democracy, but free thought itself.

A Great Awakening?

Everybody knows and feels it, although no one knows quite what to say about it. We know on some level how out of control things have gotten, but we can't even begin to imagine living any other way. The reaction is less intellectual than visceral. Life feels off. People feel stressed, behind, out of sorts, disconnected, lost. It's not just the whacked-out politics of the current presidential administration, not just the political polarization, not just job anxieties, not just the upheaval of industrialism giving way to the computer age. It's both more than that, and less.

"I think if you zoom out, this is like 1946 in a sense that we've just invented this new, powerful, and very dangerous new technology," says Harris. "We've developed a system of manipulating our own [social] system that is more powerful than the ability of even our own mind to track it."[26]

On that score, it's impossible not to be reminded of Mary Shelley, who wrote her famous *Frankenstein* in 1818 at a turbulent time—in many ways not unlike our own—that was causing romantics like her to declare a "crisis of feeling" as others before them had declared a crisis of faith. Frankenstein's monster became one of the three most enduring fictional characters of its type. But unlike Dr. Jekyll or Dracula, her monster was never named. (People mistakenly call the monster Frankenstein, but Dr. Victor Frankenstein was in fact the monster's creator.) The monster has no name because it is simply too unknowable, too unconstrained, and too wild to be tamed in that way.

Not unlike that nameless monster, Big Tech's salient feature is intangibility. These companies traffic not in widgets that we can see and touch, but in the abstractions of bits and bytes. In the long history of economic development, this is unprecedented. The wheel. The Roman aqueducts. The printing press. The key inventions of the industrial age—the previous height of technological innovation— were all immediately perceptible to the senses, and often a few of them at once. There was no mistaking an automobile, a light bulb, or a telephone. When a freight train roared by, there was no confusion about what it was, what direction it was going, whether it was carrying human passengers or livestock, and so on.

Big Tech, as vast and pervasive as it is, operates completely by stealth: silent and invisible, without shape, color, or smell. We can't see it, we don't understand it—and yet we oddly welcome wherever it might choose to go. Funny how we pride ourselves in our suspicions of salesmen, but let down all our usual defenses when some glittering new piece of technology comes calling.[27] In this, the relationship of Big Tech to its users resembles that of the retired cat burglar once played by Cary Grant to the rich and fluttery wives who must have been his victims. One imagines Grant as all charm and suaveness as he slips their diamond necklaces from their necks and nips their earrings off their ears. These ladies would be so delighted by the sheer joy of Grant's presence in their lives, they wouldn't notice that he has relieved them of all their jewels.

Part of this is due to the unique aesthetics of the technology itself. If the enduring image representing the industrial revolution was that clanging, smoke-belching locomotive roaring over the countryside, the information age is represented by the sleek, slender iPhone, probably the loveliest mass market product ever made. But wait: It's not only beautiful; it also does all this great stuff. Has anything ever felt quite so good in the hand, offered so much to the eye, *and* so enlivened the passions and quickened the mind? Canadian communications theorist and philosopher Marshall McLuhan once observed that every new wave of technology contains all the previous waves within it,[28] and so the smartphone is the telephone, the camera, the movie, the phonograph, the radio, and so much more, all in one. And all of it thanks to computer chips that forever pack more and more power into less and less space—and now, with quantum computing, will operate on hyperdrive.

A smartphone's powers are both unknowable and immensely powerful in the way that magic, by definition, is. People might have a rough understanding of how a car engine works, and if they don't, they can always pop the hood to see. But who has ever peeked under the cover of their iPhone to see what's going on inside? Who even understands how this little pocket-sized computer sends and receives photos, or summons the Internet, or lets us stream a two-hour movie? For most people this is wondrous, even magical. In fact, the transmission of the electronic signals that carry email isn't so different from the movement of sound waves through the air. Still, it's amazing that no one needs to give any of this a thought—until there's a glitch, of course.

Few things in life are so mysterious, and yet so utterly normal that we take them for granted. About the only thing that comes close is religious faith, which can likewise take hold of a person. Indeed, in a psychological and social sense, the current Big Tech fervor is oddly reminiscent of the First Great Awakening of the 1730s, which ignited the masses with transformative notions that were similarly hard to put a finger on. Just as the awakenings were the work of just a few thundering divines, most famously Jonathan Edwards, George Whitefield, and John Wesley, who sent forth gripping sermons from their pulpits (platforms, we'd call them today) that soon had people quaking all up and down the Atlantic coast, the high priests of the tech revolution number just a handful of big players: Brin and Page, Zuckerberg, Bezos, Musk, et al. Just as the ministers invoked the fear of everlasting hell if people didn't go along—and held out the promise of eternal life if they did—the exhortations of Big Tech, too, reach people at a level of consciousness located somewhere below reason—at "the bottom of the brain stem," as Tristan Harris puts it—where we have the least power to resist.

HARRIS TRIED FOR a year to shift things from his own pulpit at Google. But by 2012, he was growing concerned about the way in which the engineers at the company paid little attention to the impact of their design choices—making the phone beep or buzz with each new email, for example. Huge amounts of money and time were spent on fine-tuning details, but in Harris's view, very little was spent on asking the big question: Are we actually making people's lives better?

After a revelatory moment at (where else?) the infamous Burning Man event in the Nevada desert, he put together a 144-slide presentation and sent it to ten other Googlers (the presentation later spread to five thousand more). Entitled "A Call to Minimize Distraction & Respect Users' Attention," the deck contained statements such as "Never before in history have the decisions of a handful of designers (mostly men, white, living in SF, aged 25–35) working at 3 companies—Google, Facebook, and Apple—had so much impact on how people around the world spend their attention."[29] The presentation made waves among engineers at the middle ranks, but according to Harris, the top brass (Page himself discussed the topic with Harris) had no interest in making the business model shifts his deck suggested. There was too much money at stake.

Harris managed to parlay the slide presentation into a position (specially created, of course) as Google's "chief ethics officer." But he couldn't get traction for his ideas; there was a general sense that the company was simply delivering what users wanted; what could be so bad about that? "Nobody was trying to be evil," Harris explains. "These were simply the techniques and the business model that were standard." Nevertheless, he saw something that most people "on the inside" didn't: that we had reached a tipping point in which the interests of the tech giants and the customers they supposedly serve were no longer aligned.

"There's a reason why culture and politics are being turned inside out to become more ego driven," he says. "There's an entire army of engineers at all these firms working to get you to spend more time and money online. Their goals are not your goals." In 2015, Harris left Google after deciding that it was "impossible to change the system from within." He has since started a guerrilla movement through a nonprofit called Time Well Spent, to try to push tech companies to change their core values.[30]

"If you are alone on a Tuesday night, you'd want those thousands of engineers [at Google] who are working to keep you on the screen alone to be trying to help you not be lonely. They could be working on trying to alleviate that," says Harris. That's the key point. Developers have a choice. They can design for empathy and connection, or they can design to maximize eyeballs. "Just like the makers of cigarettes who were knowingly peddling an unhealthy and addictive product," says Harris, "they had a choice."

Many people in the Valley, and indeed many people who believe that capitalism is simply about selling people whatever they want, would say that it's not the job of Google or Facebook or Apple to decide what's moral. Their platforms simply reflect all the good and bad that is human. But plenty of people would disagree. Harris is one of them. One of the tasks of Harris's own think tank, the Center for Humane Technology, is to develop alternative business models for digital technology that are both healthier for people, and yet economically viable for the companies themselves. It would be a wise strategy for the companies to listen as they draw increasing scrutiny for everything from privacy to monopoly to the health effects of technology.

The Federal Trade Commission announced in late 2018 that they would follow the lead of European regulators and begin investigating the use of those "loot boxes" that my own son Alex found so addictive, with the aim of determining whether gaming companies, who make up an industry that is forecast to be worth $50 billion by 2022, are knowingly using gambling techniques to hook kids. "Loot boxes are now endemic in the video game industry and are present in everything from casual smartphone games to the newest high-budget video game releases," said New Hampshire senator Maggie Hassan, who called for the investigation.[31] Since then, there has been a spate of other sorts of legislation designed to shift how technology companies market to children and how content and advertising can be presented to them. Much of this has been pressed by activists like Harris, as well as others such as James P. Steyer, the CEO and founder of Common Sense Media, who was a major force behind California's new privacy legislation as well as various propositions to protect children online. In the coming years, the industry can expect only more pressure from parents, activists, and regulators to take responsibility for how their products are affecting our brains—and the brains of our children. The question is how they will respond to it.

Toward Humane Technology?

Technology firms are struggling already to get out ahead of it all, with tweaks to their products and services geared toward children. Of all the Big Tech firms, Apple has been perhaps the most receptive to criticism over the addictive properties of their products, in part because their core model doesn't depend on monetizing personal data via targeted advertising in the same way that Google and Facebook do. (Though it certainly does depend on attention: The company's App Store ten-year-anniversary press release lauded the success of games like Angry Birds and Candy Crush, which have hooked millions of people.)[32] As *The New York Times* reported in 2018, Apple devices do host apps that track users' location within the Apple orbit, but only around 200 or so, as compared with Android's 1,200.[33] And Apple has made some significant changes, prompted by pressure not just from activists like Harris, but more recently from investors—including the large hedge fund Jana Partners and the California State Teachers' Retirement System, which controls about $2 billion of Apple shares—who in 2018 sent a letter urging the company to develop new software tools that would help parents control and limit the impact of device use on their kids' mental health.[34] The company has responded by creating a new set of controls that allows users (or their parents) to track how they are using apps, and to cut the number of notifications they receive.[35]

As for Google or Facebook fundamentally changing their attention-hijacking practices, it will be an uphill battle. Google, which has been more receptive to feedback than Facebook (though that's not saying a lot), has changed some algorithms on YouTube, for example, to try to combat the problem of filter bubbles. And it has also, as mentioned earlier, considered moving children's You-Tube content onto a separate platform. But it's difficult to see them successfully shifting their entire business model to revolve less around data collection and the monetization of attention, and like any legacy company, they are reluctant to change what is already so profitable. It's likely that only a threat of regulation would prompt them to do that, and indeed, that's slowly but surely happening. In a properly functioning market, start-ups might move into this fray and disrupt the paradigm with new business models that maximize utility rather than time spent online. Some have tried, but the mo-nopoly Google and Facebook hold in their respective areas makes it very hard for innovators to gain traction.[36]

As Guillaume Chaslot, the former Googler who tried (unsuc-cessfully) to shift the nature of algorithms at YouTube to combat filter bubbles, put it to me, "There just aren't any incentives at the big companies to change business models. You need start-ups to do this. But they don't have scale to compete, and they can't get the funding to grow," since nobody will invest in competing technology because the network effects harnessed by the largest players seem too powerful to disrupt.[37] How these networks and their disruptive effects work and how they are moving throughout not just con-sumer technology, but every industry, is the topic of the next chap-ter.

CHAPTER 7

The Network Effect

Emails are the gift that keeps on giving. Facebook and Google have tried for years to brand themselves as champions of freedom, democratizers of information, and connectors of the world. But when you look at their internal email trails, you often see a different story. So it was in the winter of 2018, when British lawmakers released a trove of Facebook emails dating from 2012 to 2015 that provided a window into the duplicitousness of the company's top brass.

It should come as no surprise that any big company would be single-mindedly focused on growth. That's what capitalism—at least the kind we have in the United States and most parts of Europe at this moment in time—is all about. But what's less expected is the extent to which the tech giants have been allowed to employ anticompetitive practices to sustain and even accelerate that growth, in ways that surely would have triggered regulatory backlash had they occurred in other industries.

First is the way in which Facebook has used its size and scale to quash competitors. As the network grew, Facebook became a company with monopoly power. Like a railroad or a utility, it ran a platform that people needed access to if they wanted to reach a certain audience. It could therefore demand almost anything it wanted from those who needed that network—and its data—to develop their own businesses. And, by the same token, Facebook could *deny* anyone access to those massive amounts of user data (which is the only reason other businesses are interested in being on Facebook in the first place), for any reason.

As the 250 pages of emails and documents released by British lawmakers revealed, companies who were not considered competitive with Facebook, including Airbnb, Lyft, and Netflix, got preferred access to data, as did the Royal Bank of Canada and a number of other nontech businesses. But those companies that Facebook viewed as competition, like Vine (a Twitter-owned video app), were denied or even shut out of the company's network altogether. Indeed, after Twitter released Vine in 2013, Facebook shut off Twitter's access to Facebook friends data at Zuckerberg's behest.[1]

Meanwhile, the emails revealed that Zuckerberg discussed charging app developers for access to Facebook user data, while also forcing them to share their own user data with Facebook's network; email debates show that the company even considered restricting developer access to certain kinds of data unless the developers bought advertising on Facebook. "It's not good for us unless people also share back to Facebook and that content increases the value of our network," wrote Zuckerberg. "So, ultimately, I think the purpose of platform . . . is to increase sharing back into Facebook." In another email, his COO Sheryl Sandberg advocated the same idea. "I think the observation that we are trying to maximize sharing on Facebook, not just sharing in the world, is a critical one," said Sandberg, in a telling departure from her and Zuckerberg's public refrain about "making the world more open and connected."[2]

When it came to growing the network, it seems that nothing was off limits. And it's no wonder, because as we will learn, the network is where the value lives. Facebook *needed* to grow it, at all costs. That's why executives agreed to risk potentially bad PR so that Android apps could allow the logging of users' phone calls. It was an invasion of privacy on a new level—but it also created more data that could be mined, which increased Facebook's ability to grow.[3]

The Operating System for People's Lives

In 2011, the FTC launched an investigation into Google (this was around the same time that a variety of state agencies as well as European and Asian regulators began looking into the company's competitive practices), centered around the claim that Google had monopoly power in various markets and would use it to crush competitors if it could. The case was prompted in part by complaints brought by Yelp, the popular search service specializing in deep, hyper-local information about individual communities (like, for example, which daycare service is best according to a group of local users in Portland, or where to get the finest Thai food in Boston).

The trouble had started years before, when Yelp, which was at that point just a nascent start-up, signed a deal with Google, allowing the search giant to use some of its content, including the reviews of local services that users had posted on Yelp. This was a win-win, because it allowed Google, which at the time didn't really do local search, to have access to more specific content that lived on Yelp's site, and it increased traffic for Yelp, which needed more eyeballs. "It was better to be friends than enemies at that stage," according to Yelp cofounder Jeremy Stoppelman, who started the firm with fellow tech entrepreneur Russel Simmons.[4] Plus, they depended on Google to help them generate traffic. It was, after all, the main search highway on which most consumers began their travels, thanks to the network effect that favored the largest players.

From the beginning, Yelp and other such "vertical" search engines were in a different kind of search business than Google: They dealt in extremely precise and narrow types of content rather than the "universal" searches that were Google's bread and butter. But as local search became more and more popular, Google decided to get deeper into that business. Plus, the rise of the smartphone, as well as the growth of Amazon, was putting pressure on Google to solidify its position in the larger tech marketplace. If the company wanted to be the operating system for people's entire lives, it couldn't afford to cede any piece of the search market—big or small. So, a couple of years after inking the deal with Yelp, Google decided to add a "local search" feature that allowed its own users to review and rate local services, just as Yelp's did. Worried (and justifiably so) that Google was trying to copy his business model, Stoppelman decided not to renew their contract. Two years later, Google tried to buy Yelp for a whopping $550 million, but the deal fell apart.

That was when the gloves came off. At that point, "half of all smartphone searches had some kind of local intent behind them," says Luther Lowe, Yelp's policy director and one of the driving forces behind its efforts to get regulators to pay attention to Google's anticompetitive practices.[5] So to capture that traffic, Google, according to Lowe, began to display its own local search results above those of Yelp and other competitors, under the guise that this would create a more user-friendly interface, and actually create "better" results.[6] Of course, as reams of complicated FTC documents make quite clear, *better* was defined, quite simply, as whatever was better for Google.

"They had a massive incentive to exploit their network dominance at that point," says Lowe, "by siphoning people into their own Yelp clone." In a memo that was accidentally leaked to *The Wall Street Journal* during an open records request,[7] FTC staff noted estimates showing that creating a more equal playing field would have led to "annualized loss of $154 million" for Google on product queries. "In certain areas where Google already had existing vertical properties, such as shopping and local, Google saw a critical need to invest further and take measures to increase user traffic to those properties," notes the document.[8] As a result, mandates came down in executive meetings to find ways to fend off competition, regardless of what it meant for search quality. ("Larry thought [Google] should get more exposure," reads one footnote.)[9] And, eventually, that meant preferred placement at the top of search results. "When Google's algorithms deemed local websites, such as Yelp or CitySearch, relevant to a user's query, Google automatically returned Google Local at the top of the [result list]."[10]

As the leaked document makes clear, this "by any means necessary" approach to competition at Google originated at the highest rungs of the company. In one section, Marissa Mayer, who at that point ran Google's local, maps, and location services division, argues for calculating search results in such a way as to favor the company's own services.[11] In another, Google chief economist Hal Varian admits that "from an antitrust perspective, I'm happy to see [comScore] underestimate our share."[12]

According to Lowe, who has spent the past several years trying to prove to regulators on three different continents that Google unfairly gave preference to its own products, "The core motivation of Google is to be the middleman of all activity on the Internet." By using the power of their network and ecosystem, Google knew it could effectively banish competitors like Yelp from the Internet altogether. For a few years, says Lowe, "if you opened an Android phone and it had a Google Places app preinstalled and you clicked on it, you might open a Yelp review of a restaurant or whatever, but with no link back or attribution whatsoever."

That changed in 2011 when Yelp staff started appearing in front of state attorneys general and other regulators to tell the complicated story of how Google was handicapping competitors in areas where it wanted to grow its own business. Lowe recalls in particular one conference in Hawaii, where Yelp staff delivered a presentation so compelling that several of the AGs (some of whom have since pursued cases against Google) were forced to sit up and take notice. Google also had staff in the audience, and afterward, says Lowe, "they looked like they'd seen a ghost." Google's treatment of Yelp content improved somewhat after the conference, but by that time, says Lowe, "the damage was done, and Google had built enough of a following and collected enough of its own reviews that it didn't need Yelp as much anymore." Yelp survived, but has struggled to maintain its market share, and is, of course, a fraction of the size of Google. The network effect had done its work.

The case was ultimately dismissed, which was itself unusual given that the recommendation to bring the lawsuit came from the FTC's own bureau of competition. Many sources I've spoken with feel that the decision came down to Google's lobbying power in Washington; the company not only sent its top brass to lobby politicians directly, but also funded research favorable to its cause—some of which was done by academics such as Joshua Wright, who joined the FTC shortly before the case against Google was dropped.[13]

Kent Walker, Google's chief legal officer, played down those concerns, without specifically denying them. "With regard to lobbying, it's important to remember that the FTC professional staff, three different commissions, the Bureau of Competition, the Bureau of Economics, and General Counsel's Office all reviewed this case and all found that Google was acting primarily on behalf of consumers with regard to the innovations we were making."[14] While Walker has always struck me as a decent guy—someone who is primarily just trying to clean up whatever legal messes the company gets itself into—I didn't really buy his line. After all, it wasn't until after Renata Hesse, a former Google counsel, had been appointed the acting DOJ antitrust chief (and Larry Page had met with FTC officials), that the case finally went away.

The issues themselves, though, remain. In fact, there are new bipartisan calls to reopen the case. Meanwhile, the company is now battling EU regulators in similar, ongoing investigations into whether it has used anticompetitive practices to winnow out smaller companies in search. In 2018, for example, the company paid a whopping €4.3 billion antitrust penalty to the European Union for abusing its power in the mobile phone market: nearly double what it was charged a year earlier for favoring its shopping service over competitors' in its search results.

That case centered around two British technologists turned entrepreneurs named Adam and Shivaun Raff, a husband-and-wife team who launched an online price comparison site in the United Kingdom in 2006. The two were programming nerds who'd come up with an algorithm that was very good at particular shopping queries. (For example, what airline has the cheapest flight from Glasgow to Madrid on Tuesday? Or, what's the best vacuum cleaner with a HEPA filter?) This sounds easy, but it's not; specific, deep-search queries like this are actually much tougher to service than more universal ones. But the Raffs had cracked the code, and within forty-eight hours of launching their site, which they named Foundem, they were flooded with traffic from shoppers looking for everything from computers to appliances.

Then the traffic stopped, pretty much cold turkey, according to Shivaun Raff, who approached me in 2017 following an article I had written in the *FT* on the way in which Big Tech used the network effect to achieve monopoly power. Or, to be more specific, the traffic from Google stopped. While Foundem would rank at the top of results from other search engines—including Yahoo and MSN Search—Google would bury them way down the list. Given that research shows that users pay attention to only the top five search results,[15] Foundem was effectively banished from the Web.

The way in which Google was able to effectively banish both Yelp and Foundem from the top listings in searches, and thereby effectively stop them as competitors, has major similarities to the way other "essential facilities," like the railroads and telecommunication lines of old, were able to hold up competitors and customers alike, providing access to their networks, or not, for whatever fee they liked, or in whatever way they liked.[16] In 1900, for example, six U.S. rail companies owned or controlled 90 percent of the market for anthracite coal, resulting in high prices for buyers and massive profits for the railroads—which of course made it difficult for the independent coal companies to move product over their lines.[17] The problem was eventually rectified through a "commodities clause" that separated platforms and commerce. This kind of separation eventually made its way into other areas such as banking, preventing bank holding companies from competing with their own clients in various industries (though they sometimes got around such things via regulatory loopholes).

The Internet is, of course, the railroad of our times—an essential piece of public infrastructure over which much of the world's commerce and communication is now conducted. And the parallels between what nineteenth-century regulators called "the railroad problem" and the Internet problem of today are strikingly similar. As part of the research for this book, I cracked open a slim but surprisingly readable 1878 volume entitled *Railroads: Their Origins and Problems,* by Charles Francis Adams,[18] a former railroad executive and regulator, who lays out the rise of the railways in both Europe and the United States, and the struggle to force them to serve the public at large rather than just a handful of industrialists. In a chapter called "The Railroad Problem," Adams writes, "As events have developed themselves, it has become apparent that the recognised laws of trade operate but imperfectly at best in regulating the use made of these modern thoroughfares by those who thus both own and monopolise them."

You could, of course, retitle the same chapter "The Internet Problem" and have a good summary of where we are today. Amazon captures over one-third of all U.S. online retail spending, a figure that was recently downgraded by the company itself from previous estimates of around half. Amazon attributes this revised estimate to changes in how third-party sales are accounted for, but it has nonetheless aroused suspicion that the company is fiddling with its metrics to get ahead of regulators trying to make an antitrust case.

Google represents 88 percent of the U.S. search engine market, and 95 percent of all mobile searches. Two-thirds of all Americans are on Facebook, which, having bought Instagram and WhatsApp, now owns four of the top eight social media apps. All of these companies, as well as Apple, the world's first trillion-dollar company, have come under fire for using their enormous ecosystems to give their own products and services preference and keep competitors off their networks.[19] It's a problem inherent in both owning a platform and conducting business on it.

But the monopoly argument is one that neither Google nor any other Big Tech giant is willing to admit has merit. According to Kent Walker, the failures of competitors Yelp or Foundem had little to do with anything that the search giant did. "There's an awful lot of good econometric data that shows that the decline of various services came for a whole variety of reasons unrelated to the evolution of Google search results," he says. Not only did Google's own results get better, says Walker, but other giants like Amazon were rising. That may well be the case, but it's quite telling that the search competition was increasingly among giants (indeed, Amazon has since become Google's main rival in search) who had the power to keep smaller companies entirely out of the market.[20]

The Raffs, for their part, believe that Google was purposefully trying to shut them down. After traffic tailed off, the Raffs, who knew people within Google and were well-connected within the global tech community, began reaching out to people at the company, to no avail. It was no secret that Google had been working for years on its own Google Shopping service (formerly known as Froogle),[21] but the Raffs didn't immediately assume there was sabotage involved.

"For three and a half years, we went through various channels, official and unofficial. We just never got any meaningful response about what was happening," Shivaun Raff says. Meanwhile, they began seeing similar stories on programmers' websites, tales of entrepreneurs who'd gotten too close to the company's core business model and were effectively put out of business. Some described the all-too-familiar experience of disappearing from search results, or claimed that Google had put pressure on their customers or clients; others recounted lawsuits that took such a financial toll that they drove smaller players out of business.[22] The Raffs themselves soldiered on, eventually building up a user base of more than 2 million people, but they were draining their savings, and without the Google traffic, it was a struggle just to meet the day-to-day needs of their business. "Google is the gateway," says Shivaun. "If you are excluded from that ecosystem, you die."

Eventually, they decided to "stop being British," as Shivaun put it to me, and take their case to the regulators, which is how Foundem became the lead complainant in the European Commission's Google Search antitrust case, launched in 2009. It was led by the tough-as-nails EU competition chief, Margrethe Vestager, who eventually found against the firm in 2017. In compliance with EU law, Google was given eighteen months to figure out a way to rejigger its algorithms to eliminate bias in search. But in late 2018, the Raffs sent a letter to the commissioner, telling her that they were unpersuaded that the Google "compliance mechanism," which depended once again on its own black box algorithmic formulas, was working. "It has now been more than a year since Google introduced its auction-based 'remedy' and the harm to competition, consumers, and innovation caused by Google's illegal conduct has continued unabated," they wrote. "We therefore respectfully urge you to commence non-compliance proceedings against Google."[23]

It's quite possible that the EU will do just that in the coming months or years. But it will likely be too late for the Raffs, who have had to turn back to consulting to make their livings, and are running Foundem more as an act of defiance—and determination to win their legal battle—than to make real money. "We're not stopping," says Raff. "They are going to have to change the way they do things."

The Power of the Ecosystem

Google itself isn't worried about competition from upstarts but from other behemoths, namely Amazon, which is the only company that is really giving the search engine a run for its money these days. Many people who are shopping for products, as opposed to looking for general information, now start their searches on Amazon, which means that advertising money is also migrating to the e-tail giant. WPP, the world's largest purchaser of advertising (it shops on behalf of major companies globally), spent $300 million on "behalf of its clients on Amazon search ads last year," and about 75 percent of that was pulled from budgets that would have been spent on Google-related advertising. No wonder the Googlers are looking nervously over their shoulders to see what Jeff Bezos is doing.

All of this shines an uncomfortable light on how the supposedly decentralized Internet economy has spawned a handful of ruthless oligopolies that have begun to use their power to undermine start-up growth, job creation, and labor markets. Over the past two decades, more than 75 percent of U.S. industries have seen an increase in concentration of both wealth and influence. If you compare the numbers with the post–World War II period, when U.S. growth was strongest, the contrast is striking. In 1954, the top sixty companies accounted for less than 20 percent of U.S. GDP, according to the Brookings Institution. Today, the top twenty companies make up more than 20 percent.[24]

Why is this happening? One reason, of course, is global competition, which has put more pressure on U.S. businesses, and pushed companies away from the more equitable postwar pie sharing between workers, corporations, and local communities. Another is the shift in antitrust law. But another, less explored, reason is the network effect that goes along with the platform technology business model.[25] Concentration is happening everywhere, but it's most pronounced within the information economy. (According to the McKinsey Global Institute, industries such as tech, pharma, and finance, which are based on data and intellectual property that can be monopolized and moved anywhere around the world, are the most prone to concentration.)[26]

It is the economic and political challenge of our time. Jason Furman, the former head of the Council of Economic Advisers, believes concentration is creating barriers to entry in many key markets.[27]

Academic David Autor has linked the corporate consolidation to workers making less.[28] New research from the McKinsey Global Institute has found the same, and has noted in particular the way in which technology has driven down the labor share of the overall economic pie.[29] There is also evidence that a small group of "superstar" companies are pulling way ahead of others, not only in terms of profits but also productivity.[30] In other words, the biggest companies, particularly in the most digitally connected parts of the economy (tech, finance, and media), are incredibly productive. Everyone else, not so much. The upshot is that economic growth as a whole has suffered.[31]

The Goliath versus Goliath idea was recently given some heft by a 2018 McKinsey Global Institute analysis of nearly six thousand of the world's largest public and private companies, each with annual revenues greater than $1 billion and which together make up 65 percent of global corporate pre-tax earnings. Among this group, the top 10 percent (the "superstar" companies) take 80 percent of economic profits—defined as a company's invested capital multiplied by its returns above the cost of that capital. The top 1 percent alone take 36 percent of the pie.[32] We already know who some of the top 10 percent are—they include high-margin Big Tech companies (Facebook, Apple, Amazon, and Google), as well as a number of others who have been able to exploit the value of intangible assets such as software, data, patents, and brands (these would include not just technology firms but also a good number of financial, biotech, and pharmaceutical companies). We also know that the network effect allows such companies to capture markets quickly and at scale, giving them what's known in the start-up world as the "first scaler advantages."

This process is aided by the fact that we have shifted from a "tangible" economy, based on physical goods, to one based more on intangibles—namely intellectual property, ideas, and data. British academics Jonathan Haskel and Stian Westlake, who lay out the case for this in their excellent book, *Capitalism Without Capital,* believe that this shift upends the usual rules of economic gravity. Google and Facebook don't need to build more factories, invest in more raw materials, or staff more assembly lines in order to capture market share, which is why they have the ability to grow much faster than corporate giants of the past. In today's economy, the losers tend to own more things—tangible assets such as factories and equipment—whereas the winners are concerned with leveraging intangible ones.

The network effect is at the center of this shift. Whether it's made up of Twitter users, Uber drivers, Airbnb hosts, or Instagram influencers, the network is worth far more than the value of any single node within it. The key point is that users beget users, which allows the players who can grab the most market share quickly to dominate entire industries seemingly overnight. This is not unique to Google, as we've seen. But it is much easier these days to grab market share if you are big, and can leverage data and intellectual property across networks. These intangible assets can scale far, far faster and further than the products and services of old.[33] Networked businesses are case studies in how what goes big, goes bigger still.

But as Varian and Carl Shapiro acknowledge, there's a dark side to the feedback loop: "Positive feedback [within platforms] also makes the weak get weaker." In other words, even superstars of the networked era can fall, though typically to each other, rather than to upstarts.[34]

JUST LOOK AT the car industry for a sense of how profound the shift will be for traditional firms. Currently, about 90 percent of the value of an automobile lives in the hardware. But as autonomous driving and digital apps become a bigger deal, that ratio is expected to shift dramatically. Morgan Stanley predicts that in autonomous vehicles, 40 percent of the value of an automobile will come from hardware, 40 percent from software, and 20 percent from the content that streams into the vehicle.[35] That would include things like games, advertisements, and news enabled by the software. This shift is partly driven—no pun intended—by the fact that millennials want their cars souped-up with all their favorite apps. But it also reflects another idea: When you are in an autonomous vehicle, brand identity disappears.

"If you take away control of the steering wheel, consumers are much less likely to care what type of car they are in," says Nick Johnson, principal at the consultancy Applico, who has advised major automakers on the shift. Johnson is also the author of *Modern Monopolies,* which looks at the effects that the Silicon Valley giants are having on other firms and industries. In this world, a car is no longer something you touch and feel or "wear" like a luxurious garment, but something you use—like a phone. And if that is the case, it is the software and apps that develop around the software platform that really matter, not the plastic and metal shell they live in, as phone makers Nokia and BlackBerry can attest. Indeed, a recent survey by carmaker BMW revealed that 73 percent of people said they would exchange one brand of car for another if they could bring their digital lives into the new car.

This is the challenge facing GM, a company that was attacked in 2018 by Donald Trump and labor officials alike for laying off or buying out 14,300 autoworkers (part of that as a result of shuttering five factories in the United States and Canada). Both the U.S. president and the unions focused on arguments about sending jobs to China and Mexico. But the biggest challenge GM is facing isn't one of labor costs, outsourcing, or steel tariffs. It is the question of whether it will be able to continue to own a large share of the economic value of the automobile industry in the networked era in which the car is becoming a smart device. It's a question that faces any number of industries, from retail (which has already been decimated by Amazon) to healthcare (under competitive threat from both Amazon and Google), finance (which is under threat from fintech, the merging of tech platform technology and banking), manufacturing, and so on.

The businesses that have the best technology in automotive software and apps are, unsurprisingly, technology companies such as Google, Apple, and China's Baidu, all of which are pouring money into autonomous vehicle technology and platforms to support it. Right now, motorists can mainly stream music, GPS information, and whatever other data they can access on their phone via such systems. But once the platforms are embedded more deeply in vehicles, customers will be able to tap into everything from fluid levels and engine temperatures to safety information. Those are all currently the domain of carmakers. Monetizing all that data, via new products and services, is the big prize.

How should companies think about competing in this world? The case study on what *not* to do comes from Nokia. Remember the once-mighty Finnish phone maker? You may have used one of its brick-like handsets to type some of your first text messages back in the 1990s. Then Apple's iPhone came along, followed by Google's Android operating system. Both companies offered not just snazzy products but successful platforms for developers. Ecosystems of apps grew around them, while Nokia's Symbian operating system became, by comparison, hopelessly passé. By 2011, Nokia was in free fall, never to recover.[36]

As Symbian head Jo Harlow told the *Financial Times* at the time, the company simply hadn't been quick enough to make the shift from being "device-led to software-led."[37] True enough. But the larger problem was that, in a deeper way, Nokia, like many before and after, failed to see that not only would the vast majority of value migrate from hardware to software, but specifically to *the platform on which that software would operate*. The network effect of developers and users around those platforms is what would create value—much more so than the product itself.

Carmakers are not standing still. GM's chief executive Mary Barra has for years referred to GM as a technology company, one increasingly dependent more on data than on steel or manpower. In 2018, she explicitly tied the big GM job cuts to a shift in resources toward electric cars and autonomous vehicle development. There are some nascent industry partnerships, such as Ford's SmartDeviceLink Consortium, an open source community working via a standard set of protocols to connect smartphone applications to vehicles. But no big carmaker has shown itself able or willing to create the platform ecosystems that big technology companies create. This is a problem, because the network effect really kicks in when a company controls 30 or 40 percent of a given market, which means that the major auto companies of the world would need to team up in order to achieve such a share. Certainly, it would be a big shift for a company to think about its most aggressive competitors as collaborators. Yet it may be the only choice they have. Developing an ecosystem and owning the software and data within it will be the key to success not just in the car business, but in many industries.

Neoliberalism on Steroids

As powerful as the network effect is, to understand the seemingly unstoppable growth of the platform companies like Google or Facebook, you also have to look at how much the politics of Silicon Valley changed between the era of hippie idealism represented by Steve Jobs, and the libertarian epoch of Peter Thiel and his ilk. "It was a titanic shift," says Roger McNamee, who has worked in tech for more than forty years. "While the rank and file in Silicon Valley is liberal, the top people at the top firms tend to believe that greed is good."

How could they not? Ever since the 1980s, most of American business has been subscribing to the trickle-down "markets know best" doctrine popularized by the so-called Chicago School of economics. The Internet platforms in particular have benefited enormously from the Chicago School's antitrust philosophy, which maintains that as long as products are cheap or free, there's no monopoly issue. As McNamee outlines in his own book, *Zucked,* "Google leveraged its dominant market position in search to build giant businesses in email, photos, maps, videos, productivity applications, and a variety of other apps. In most cases, Google was able to transfer the benefits of monopoly power from an existing business to a nascent one." When you go back to the economic history books, this should come as no surprise—monopoly power was a central feature, even an aim, of those who invented the economics of information technology.

How the Big Get Bigger

The neoliberal politics of Silicon Valley are reflected in the work of the economist Hal Varian, who joined Google as a consultant in 2002. Eric Schmidt had run into Varian back in 2001 at the Aspen Institute, one of those places where tech titans and their admirers gather to discuss Big Ideas. Schmidt informed Varian that the company had an auction model that "might make a little money" and asked if he would come and help the company perfect it.[38] Varian, who'd been dean of the Berkeley School of Information, was one of the top economists studying data markets at the time. He had co-written an influential book entitled *Information Rules: A Strategic Guide to the Network Economy,* and would eventually have a rule named after himself—a sort of trickle-down theory for the digital age. The Varian rule posited, incorrectly, that everything rich people had today, the middle classes and eventually the working classes would have tomorrow, thanks to the price-crunching effects of technology. (Big Tech critic Evgeny Morozov later rephrased it in perhaps a more factually accurate way: "Luxury is already here, it's just not very evenly distributed.")

Right around that time, he gave a series of lectures and wrote a number of papers that laid out some of the key ideas emerging from the burgeoning field of data economics, ideas that make it hard to believe that the people at the top of today's platform technology firms didn't understand the far-reaching and potentially disturbing effects their innovations might have on our economy, our politics, and our society.

Varian, like most economists, believed in Chicago School theory, and layered ideas about the network effect and the power of big data on top of that intellectual framework. He understood that companies who can harness the network effect "have significant market power," particularly since the data they acquire "allows for fine-grained observation and analysis of consumer behavior."[39] As such companies build scale, they acquire a kind of monopoly power based on that relationship. As Varian puts it, "An extended relationship allows the seller to understand 'their' consumers' purchasing habits and needs better than potential competitors. Amazon's personalized recommendation service works well for me, since I have bought books there in the past. A new seller would not have this extensive experience with my purchase history, and would therefore offer me inferior service."[40] (Particularly if that seller can't get a leg up on the dominant platform, as was the case in the Google-Foundem conflict.)

Varian eventually became chief economist at Google, where he quickly built his reputation as a practical diviner of this new economics of information. He hired an entire team of "econometricians," who combined neoliberal theory, mathematical ideas, and data to help Google make as much money as possible, then went to work helping Page, Brin, and Schmidt develop more efficient auction algorithms and build the auction model that became such a gold mine for Google.

One of his tasks was to analyze the signal in the noise of all the data that Google was gathering. He brought the new efficiencies of data economics to resource allocation at the company itself, developing an auction model that calculated and allocated internal computing power as sharply as any Wall Street trading scheme. (The paper that came out of that experience was titled "Using a Market Economy to Provision Compute Resources Across Planet-wide Clusters.")[41] Predictably, his theories of the new data economics tended to favor his employer. As Shoshana Zuboff has written, in the sort of surveillance capitalism practiced by Google and other Big Tech firms, "Contract and the rule of law are supplanted by the rewards and punishments of a new kind of invisible hand,"[42] the algorithmic hand of Silicon Valley.

Varian and his team were unique, and foreshadowed an era in which most big companies would hire data scientists and data economists in great numbers. The existing laws that governed commerce were, like most laws in the view of Big Tech, made to be broken.

Stewards of Trust?

To be fair, pioneers like Varian have acknowledged a number of downsides of this new networked business model being pursued by Google and numerous other Silicon Valley giants, even the big one: privacy. Amazingly, in 2011, he admitted that, as a user, he would not want the platforms to share personal information with third parties without his consent.[43] He concludes, however, that this problem isn't much of a risk; the sale of information to third parties without consumer consent wouldn't be economically efficient, since it would breach trust.

It was a position only slightly more self-aware than that of his boss, Eric Schmidt, when asked, in a 2009 CNBC documentary, *Inside the Mind of Google*, about whether people should trust Google with their most personal secrets. Schmidt replied, "I think judgment matters. If you have something that you don't want anyone to know, maybe you shouldn't be doing it in the first place." Translation: Your privacy isn't our problem.

According to Nobel Prize–winning economist Paul Romer, much of our willingness to trade our right to privacy for the sleekest new iPhone model has to do with the fact that "there are tremendous information asymmetries in these markets. Do both parties understand enough to know whether the transaction taking place is in their mutual interest?" he asked, rhetorically. Romer (like me) would argue that they do not; he believes that the complexity of today's data markets "means that notions like 'consent' [to long and complex disclosures about how your data might be used by platform companies] have become meaningless." The differences in what either party knows simply undermine the fair functioning of the market itself. "I've been in these discussions with people like Hal Varian, and I get more and more frustrated," he told me in 2018, shortly after winning the Nobel. "There's a dishonesty about giving people something that's eighteen thousand words long and expecting them to read and understand it," Romer says.

But what's the solution? For starters, says Romer, we should stop using the word *privacy*. "It doesn't really exist anymore," he says. We should focus more on transparency and clarity. "If nobody—let's call that fewer than five percent of users—can get an even partial understanding" of the terms of a transaction, then Romer says companies simply shouldn't be doing them. What's more, "we should put the burden of proof on the companies themselves," rather than allowing them to circumvent responsibility via "phony disclosures."[44]

Some companies, such as Apple and even IBM, which is still very much a key player in the technology world, are finally waking up to the idea that protecting user privacy is a competitive advantage.[45] Apple, for example, has rolled out a new website to better showcase privacy features that it believes differentiate the company from its competitors, including an algorithmic function whereby search data is stored within individual devices rather than in the "cloud," thus giving users more control over what the company can see.

Apple is also touting a technique known as "differential privacy," which allows the company to gain insights into what users are doing, while preserving a certain amount of privacy by encrypting the data before it leaves a user's device, in such a way that Apple can't associate the data it receives with any particular user. The data is used to improve the devices and services that are sold within the Apple ecosystem, rather than to send customers hyper-targeted ads from other businesses that they had no idea were getting their data to begin with. That is, again, quite a departure from the Google/ Facebook approach. Does this solve all the problems? No. But on the other hand, Apple's business model doesn't lend itself to influencing an election like Russia attempted to do in the United States via Facebook. It's also refreshing to hear Tim Cook say that he believes "privacy is a fundamental human right."

Ginni Rometty, IBM's chief executive, has also announced a new set of principles and practices around data aimed at increasing trust in Big Tech. These include a pledge not to keep any proprietary data in their servers for more than a specified contract period, never to turn over client data to any government surveillance program in any country, and a promise that clients will own not only the rights to their end data, but to any algorithmic "learning" from it.

"We're entering an era in which data can be used to solve all sorts of the most pressing problems, but only if there's trust in how that data has been handled," Rometty told me in a phone interview in 2017. "We see ourselves as stewards of clients' data. And we don't need to be regulated to do the right thing. We've been doing the right thing for a hundred years."

To be clear, neither Apple nor IBM are perfect stewards of trust— both own apps that track data. And IBM has the advantage of dealing more with other businesses and governments than with consumers directly. But they also show it's possible for companies to start addressing such issues.

How many companies will be able to successfully square the growing concerns over privacy, transparency, and monopoly power? What effect will these have on the markets, consumers, and citizens? How should these new platform giants, which grow to scale in a world in which the usual economic laws of gravity cease to function, be regulated? These are pressing questions, because as we've already seen, not just from Amazon or Google, but from start-ups like Airbnb and Uber as well, digital giants can come from out of nowhere and disrupt incumbents, consumers, workers, and even entire cities in one fell swoop, at a pace that would have once been unthinkable.

CHAPTER 8

The Uberization of Everything

February 2017 wasn't a good month for former Uber CEO Travis Kalanick. The ubiquitous ride-hailing business he founded had been drawing criticism from municipal lawmakers and union activists—particularly in large cities like New York and San Francisco—for years, but their PR crisis reached a boiling point following a series of scandals that started with a blog post from a former engineer, Susan Fowler, alleging harassment and rampant sexism at the company. That news went viral in the same month that Waymo, an autonomous vehicle unit owned by Google's parent company, Alphabet, filed a federal lawsuit against the ridesharing company alleging that a software engineer had stolen its trade secrets and taken them to Uber, which is developing its own autonomous vehicles.

This was followed only five days later by a shocking video showing the CEO himself blowing up at an Uber driver who deigned to complain about the company's payment system.[1] Uber's own dashcam recorded the interaction, in which the driver claimed to have gone bankrupt after investing $97,000 in a high-end car in order to drive for uberBLACK, only to find that rates began falling and the service was being dropped in favor of cheaper cars. In the video, an agitated Kalanick says, "You know what? Some people don't like to take responsibility for their own shit. They blame everything in their life on somebody else. Good luck!" To which the driver replies, "Good luck to you, too. I know that you aren't going to go far."[2]

Not in that car, anyway, and as it turned out, not as CEO of Uber, either. Especially not after the emergence of more sexual harassment revelations, resignations of key executives, and Department of Justice investigations into whether Uber used software to evade municipal authorities as it tried to build its network in various cities. In June 2017, Kalanick announced he was taking a leave of absence.[3] Later that month, with the company under investigation for multiple issues, fielding numerous civil lawsuits in which they were accused of fostering unsafe conditions for drivers, and dealing with data breaches and a massive #DeleteUber campaign, Kalanick was finally pushed out as CEO by investors. He was eventually replaced by Dara Khosrowshahi, an Iranian American former banker and protégé of Barry Diller's, who had formerly run the Diller-owned online travel firm Expedia. Diller tried to talk him out of the Uber job. "I said, 'Oh my God, Dara, you must be out of your mind. That's a very dangerous place,'" said Diller in a 2018 *New Yorker* article on the company.[4]

It was also a rich one—on paper, at least. Despite all the fiascos, Uber boasted the highest pre-IPO valuation for any company ever in history—a whopping $70 billion, despite the fact that the company had posted $1 billion–plus losses each quarter in 2018, and had yet to make a profit.[5] But the ousting of Kalanick didn't change the fundamental business model of Uber, a company that had built its reputation and its scale—like so many Silicon Valley "unicorns," by focusing on growth over profit. The problem is that when companies go public, investors generally want to see some profit. Uber's IPO on May 10, 2019, one of the most anticipated in history, was a dismal failure. The shares fell 19 percent in the first days of trading, leaving investors who'd put billions into the company underwater. Part of the problem was that Uber had simply gotten too big and fat before going to market—many institutional investors who'd already bought into the company when it was private had no reason to purchase more shares. But the poor offering was also a mark of what many investors feel is a market top in tech stocks that simply don't make any money, despite their size and disruptive power.[6]

"Always be hustlin'"

Like most people, I first became familiar with Uber as a user of its services, and like most people, I marveled at the speed with which Uber vehicles took over my Brooklyn neighborhood, as well as much of New York and most other major metropolitan areas, provoking delight from riders who love the convenience, but also ire from public officials and citizens who complained that the service has wildly increased traffic and congestion. In 2015, I went much deeper into Uberland, as I had the chance to spend a good deal of time with Kalanick himself, who was on the short list to be named *Time* magazine's Person of the Year (I was an assistant managing editor at the magazine at that time).[7]

It took me quite a while to persuade Kalanick's then–PR executive, Rachel Whetstone, a tough and cagey Briton who'd formerly been at Google and would go on to Facebook, that it was a good idea for the company to allow me into Uber's inner sanctum for a few days. She was worried about the exposure, and with good reason: Though Kalanick had yet to self-combust, he was clearly a loose cannon. But like so many high-flying business types, he was seduced by the possibility of seeing his face within the red border of a *Time* cover. And so I was given permission to follow him around and get the material I needed to write an in-depth profile, as Whetstone hovered nearby, trying to stage-manage Kalanick's self-presentation.

He talked up his love of Alexander Hamilton, whose image he used as his own avatar on the Uber site. "Hamilton is my favorite political entrepreneur," said Kalanick of the tough, self-made—some might say self-serving—former secretary of the Treasury who helped establish the country's financial system despite vicious opposition. "Hamilton could see the future. But he also understood how to connect it to the practical reality on the ground. He was a great orator, too. Maybe too good. Maybe he spoke too much."

I would have said the same of Kalanick. The most telling moments during my time with him highlight the ambivalence many of us feel about the company. The first was during a meeting at the Uber headquarters in Boston, attended by a couple dozen of the company's full-time employees, most of whom were young, elite college–educated, hoodie-wearing techies who looked up to Travis like a god. The energy in the room, the excitement to simply be in the same space with the uber-Bro, was visceral. We all enjoyed the high-end snacks in the company canteen while Kalanick was peppered with questions about his career history, Uber's new ventures into self-driving cars, and whether the company would ever consider augmenting its already hefty pay and benefits (self-driving-car engineers in the Valley can make around $2 million) by handing out perks like subsidized MBAs, as other larger tech firms do. "Oooh, it's getting hot in here," he quipped, to laughs and earnest nods of agreement around the room.

But in another session, a somewhat different view of the company emerged. Uber had rented a large auditorium space near the waterfront, and was feting a carefully chosen group of top revenue-generating drivers. Though officially considered contractors, rather than employees, these people embodied the image that Uber wanted to portray to the world of a company that was boldly reinventing work by offering anyone with a valid driver's license and a vehicle the opportunity to become an "entrepreneur" with flexibility, control, and the ability to earn as they liked in between other commitments. (There were, in attendance, single moms who drove to earn school fees, and part-time college students paying for their educations behind the wheel.)

Yet there was, even in this carefully curated group, an undercurrent of discontent. In the Q and A session with Kalanick, who appeared much less comfortable in this crowd, a familiar question came up: When will the company go public—and will contractors share in the wealth? Kalanick paused, and appeared to be choosing his words carefully. "It's something that's on our minds," he hedged. "We have to be careful from a regulatory standpoint. There's a lot of bureaucracy in being a public company." His voice trailed off, perhaps because he knew that offering shares to drivers would support the case of those fighting to elevate them to the status of employees—and therefore deserving of things like overtime pay, minimum wage protection, and health benefits—something that Uber has spent plenty of time and money trying to avoid.

Later, things got even more awkward, as he attempted to justify the no-tips rule Uber had at the time by saying that industries that allow tipping tend to underpay employees. This didn't go over well (despite being backed up by empirical proof), most likely because as Uber has grown, profit margins for drivers have been compressed. "That's ridiculous," muttered one middle-aged female driver, clearly underwhelmed by Kalanick's presentation. Others sitting near me grumbled in agreement. The CEO was quickly shuttled offstage, while the drivers and their families were placated with free pizza and popcorn.

Uber allows tips now, and it also awarded a limited number of stock options to veteran drivers.[8] Khosrowshahi, who is clearly a cooler head than the company's founder, has tried to address the various cultural problems he inherited, though others have flared up on his watch (most notably the death of an Arizona woman hit by an Uber self-driving car that slammed into her at forty miles per hour as she was walking across the street). Undoubtedly, he's made improvements, but he hasn't really changed the fundamental business model of the company, which is to move fast and break things—

such as established city transportation infrastructures by displacing traditional cabbies with cheaper, freelance drivers. And move fast it does; since 2010, Uber has shot from being a two-car operation in San Francisco to employing (though the company doesn't like that word) 3 million active drivers around the world.[9]

Kalanick, whose unofficial motto was "Always be hustlin'," has been called a visionary, a disrupter, a genius, and a jerk. One thing is certain: His company is unlike anything the world has seen before. Not only has Uber become a verb, as Google did, it has created an industry shorthand, sending countless entrepreneurs into boardrooms to pitch the "Uber of . . ." The company's own ambition ranges from autonomous vehicles to hovercrafts. (Its flying cars are projected to hit the skies of Los Angeles, Dallas, and Dubai by 2020.) In France, Uber can already get you a helicopter. In San Francisco, Uber Eats will bring takeout to your door in under ten minutes. As Kalanick once put it to me, rather open-endedly, "If something is moving from one place to another in a city—that's our jam."

But Uber is disrupting more than just transportation; it's rewriting the contract between employees and labor. Over the past several years, it has cemented its role as the most prolific and pugnacious among companies shaping the "gig economy," including Airbnb, TaskRabbit, and dozens more. They are all emblematic of accelerating shifts in the way we work: 24/7, directed by technology, and without many of the traditional protections and benefits enjoyed by the middle class. On the one hand, there is something magical about the way these companies allow people to monetize resources they already possess—a home, a car, their free time. On the other, this model is a slippery slope that, some argue, ends with workers being taken advantage of. Many experts believe that the rise of the gig economy is a key reason for stagnating wages, as it has accentuated the power imbalance between workers and companies that has been increasing for the past forty years or so with the decline of unions, and the deregulation of industry in general.

The Plight of the Gig Worker

"Gig work" seems to have reached a new apex with the rise of companies like Uber. Consider the typical non-medallion taxi driver in New York, who might work for three or more companies at once: Uber, Lyft, and perhaps even an unlicensed cab firm. There is some truth to the claim that such people are essentially entrepreneurs, with all the freedom that working for themselves entails. With Uber, drivers set their own hours and are in a sense their own boss, something Kalanick always lauded as highly empowering. "There is a core independence and dignity you get when you control your own time," he told me in 2015. Fair enough. But that's about all Uber drivers are in control of. They have no say in the company's pricing, which changes regularly depending on the level of demand and often means lowering rates to get more people into Uber cars. That varies based on the algorithm; according to my own anecdotal interviews with drivers in NYC, it has been decreasing as Uber has built its market share, and is around 20 percent now, as opposed to roughly 30 percent for the local independent cab services that some people in the neighborhood still use.

Uber touts its drivers as "free and independent" contractors, yet thanks to its automated algorithmic management system, the company is able to control how they work and penalize them when their behaviors deviate from what might be most profitable—for Uber.[10] Using artificial intelligence, Uber is able to identify a class of consumers that might be willing to pay more than others for rides, depending on their zip codes. Uber can then pocket that extra take without giving more to drivers; the pay of the worker can be fundamentally decoupled from what passengers pay. Moreover, because Uber self-identifies as a technology company rather than a transportation company, it avoids complying with protections like the Americans with Disabilities Act, that would normally apply to this type of work. In her book *Uberland,* the social scientist Alex Rosenblat rode five thousand miles with numerous Uber drivers in twenty-five cities across the United States and Canada. She found that, not surprisingly, while Uber itself took most of the upside of the business, drivers were often left to bear the cost and the downsides of the disruptive technology on their own.

Lyft, Uber's biggest competitor, has always been known as the kinder, gentler ridesharing company, in part because its CEO Logan Green has been more inclined to discuss the downsides of the sharing economy in a thoughtful and open way (that and the fact that he hasn't been caught on a dashcam screaming at his own drivers). Green is, for example, concerned about the potential mass displacement of drivers in the United States (which represents the largest single category of work for men with a high school degree or less) by autonomous vehicles. Drivers themselves have reported being able to make more money on Lyft relative to Uber, and have higher levels of job satisfaction. (Lyft was first to allow tipping to drivers.)[11] Unfortunately, these things fall at the margins. At the end of the day, the business models of the two companies are almost exactly identical; both create tremendously asymmetrical relationships between companies and workers in ways that make the latter ever more insecure. This speaks to the fact that the problems with sharing-economy companies are less about the CEO than the fundamental business model.

Algorithmic Disruption of Work

Management by algorithm—which has become a fact of life not just for taxi drivers but for any number of other workers, from Starbucks baristas who can't count on a predictable weekly schedule to delivery people who have their contracts terminated if they don't want to take jobs in certain areas at certain times—may be new, but its by-products are typical of what technology has historically done to labor. From English textile workers to travel agents, new technology destroys job categories as fast as it creates them. History has shown that in the end technology is always a net job creator; the question is how long the creative destruction lasts. Today it seems to be happening faster than our political and social systems can handle it. The depth and breadth of change being effected by the gig economy is unprecedented, and while the sheer number of workers that labor solely in the gig economy relative to the traditional economy isn't yet as high as some academics once predicted it would be,[12] the changes are still happening in nearly every industry, across pretty much every geography.

What happens when everyone is, to a greater or lesser extent, a freelancer? What happens when everyone has to have some kind of a side hustle, because a single job isn't secure enough anymore? That's one of the big existential worries that Uber creates in many people, even while they, as customers, enjoy the huge convenience and cost savings it provides.

Companies are increasingly boasting about how they want their employees to act as entrepreneurs—while neglecting to mention that what they really mean is they want employees to work hard, 24/7, without necessarily rewarding them like entrepreneurs, say with a piece of equity or a performance-based salary. Taken to a logical conclusion, it's hard to imagine why any number of jobs *couldn't* be Uberized. Many have been already, from handymen to radiologists. But with everyone working on demand, with no safety net, constantly graded up or down, the labor market starts to feel exhaustingly Darwinian. "This is what has people so agitated about Uber," says John Battelle, who helped launch *Wired* magazine and now runs a conference and events business called NewCo. "It's not a tech story, it's a social story—it's about how we are going to adapt to new possibilities. It's about what the social compact between corporations, government, and society is going to be."

This is a problem that affects not just sharing-economy companies and their workers, but a host of others, on- and offline, that use technology to monitor and control labor in ever more invasive ways. Amazon is well-known for its atrocious treatment of workers in its warehouses, which were included on the National Council for

Occupational Safety and Health's list of most dangerous places to work in the United States in 2018. Many Amazon workers report higher than average levels of stress and health problems as a result of constant digital monitoring.[13] An investigation by *The Guardian* found accident and injury reports to be common. In one case, *The Guardian* reports, an injured worker was fired before the company would authorize medical treatment. Other injured workers were reportedly denied workers' comp, or had their medical leave cut short—the predictable result of a management style that treats workers less like humans than robots.[14]

I once interviewed a well-known AI scientist, Vivienne Ming, who was offered a job as chief scientist by Amazon, one that she turned down for exactly these reasons. Jeff Bezos apparently told Ming he wanted to hire her with the aim of running real-time experiments on how "technology could make people's lives better," she recalled. "What I decided in the end is that Jeff and I had different definitions of *better* . . . pretty profoundly different." How so? "I'll give an example," Ming said, and proceeded to tell me about a patent the company had recently taken out on a little wristband that factory workers would wear, and, if they started reaching for the wrong package, it would buzz. "I'm thinking, I would never want to build that sort of thing!"[15]

Even Starbucks, a company that's been lauded for its treatment of workers (giving part-timers things like health insurance and offering to pay for online college education for all employees), has been dinged for its use of algorithmic scheduling software, which can wreak havoc on lives by forcing workers to be on call whenever store traffic grows, rather than being given set weekly or monthly schedules that they can work their lives around. After *New York Times* reporter Jodi Kantor did a front-page exposé about the topic in 2014, then-chairman Howard Schultz was forced to apologize and promise to clean up the company's scheduling system.[16] Yet, at Starbucks and, alas, at most other retailers, algorithmic scheduling has become the norm—just like "surge" pricing at Uber or Lyft.

Clearly, the advent of the high-tech gig economy means different things to different kinds of workers. For the Uber driver or the delivery person, it may feel like a kind of neo-serfdom. They get no pension, health insurance, or worker-rights protection, and work at the mercy of metrics. Many of the drivers profiled in Rosenblat's book struggle to make much more than minimum wage, after paying for their car, their gas, maintenance, self-employment taxes, and so on. Certainly, in my own interviews with Uber drivers, I've found that most see a tight trade-off between the benefits of their theoretical freedom and the fact that always-on technology can actually mean less flexibility than they might have in a higher-quality job. Many of the best-paying rides come in places and at times that may be inconvenient or stressful for them, and if they don't accept they don't get paid. Certainly, most don't share in the equity value of the company that they have helped create.

The result is that for a huge number of low-level workers (who represent the bulk of the gig economy), "you've got a labor market that looks increasingly like a feudal agricultural hiring fair in which the lord shows up and says, 'I'll take you, and you, and you today,'" says Adair Turner, chairman of the Institute for New Economic Thinking, one of many nonprofit groups studying the effect of companies like Uber on local economics.

Turner's conclusion, which mirrors that of a growing number of economists, is that the gig economy reduces friction in labor markets, meaning it solves a real need and creates convenience, but it also creates fragmentation that tends to work better for employers, who can leverage superior technology and information, than for workers. The fact that all the data is owned by Uber, and not the driver, and that drivers can't see any of it, also creates a huge information asymmetry between workers and the company, as a study done by another nonprofit group, the Data & Society Research Institute, found.

Drivers risk "deactivation" for canceling unprofitable fares and absorb the risk of unknown fares, "even though Uber promotes the idea that they are entrepreneurs who are knowingly investing in such risk." These "entrepreneurial consumers,"[17] as the Federal Trade Commission has described Uber drivers themselves (buying into the company's own language, which would designate drivers themselves as consumers of Uber's value rather than producers of it), have no access to the wealth of data on consumers that allow the company itself to make such incredible profits. As Rosenblat points out, this asymmetry is similar to that enjoyed by other Big Tech firms, like Amazon, who can steer customers to more costly products through rankings, or Google, which promotes itself as a neutral arbiter of information, even as PageRank's algorithms remain inside the black box, with any biases that might be present known only to the company itself.[18]

These strategies allow only "the fantasy that there are no more issues of power in the workplace," said AFL-CIO policy director Damon Silvers in a Harvard Business School podcast on the future of work. "In reality, companies like Uber know more about their employees, and have a tighter grip on their behavior, than any steel or auto company ever did. In the absence of workers having collective power, digital technology, AI, and cheap surveillance technology will combine to make information advantages that accrue to employers . . . at a scale and intensity we've never seen before."[19]

Superstars Take All

There's no question that low-level gig workers—from handymen to yoga instructors to childcare providers—get the short end of the stick in the digital economy. But for highly educated professionals, the digital gig economy is pure upside: a way to earn more money in less time, in ever more flexible ways. Consider the life of a freelance management consultant. He or she may charge $10,000 a day per client, using cloud computing, smartphones, social networking platforms, and video conferencing to work anywhere, anytime, which makes it easy to earn a high six- or even seven-figure salary.

Those same technologies reduce operational costs to pretty close to zero, given that the price of a virtual assistant based in India is negligible for this new cadre of high-end freelance worker, and the fact that he or she can work from home, or cheaply rent work spaces via membership schemes in companies like WeWork.

The digital gig economy, it turns out, is no less bifurcated than the analog one. This is concerning, given that a spate of new research by various organizations, from McKinsey to the Organization for Economic Co-operation and Development, points to the fact that in the next ten to twenty years, the number of people working as freelancers, independent contractors, or part-time for multiple employers will increase dramatically. In the United States, 35 percent of the labor force is already working this way. If "freelance nation" is the future, the divides in this new world will only further exacerbate the winner-takes-all trend that has driven the polarized politics of the moment.

The digital economy more broadly has already widened the gap between the haves and the have-nots, the winners being those who have the ability to access, control, and leverage technology—which is, in and of itself, connected to education, or, in other words, to money and class. As Harvard academics Claudia Goldin and Lawrence Katz outlined in their book *The Race Between Education and Technology*, technological advancements lift all boats only when people have the skills and access to utilize those advancements.[20]

The networking platforms and software of this new digital economy are resulting in cheaper prices for consumers, cost reductions for employers, and higher wages for the most skilled and educated workers, who can do more highly paid work in less time. But they have also contributed to the concentration of wealth in fewer hands, in part because there is a large body of less-educated people who are left at the mercy of technology—and those who leverage it.[21]

"Think of, say, how a top surgeon using cutting-edge video conferencing technology might now be able to do more consultations in many different countries with a wide variety of clients," says James

Manyika, director of the McKinsey Global Institute. "Compare that to a retail service worker whose life has been made chaotic by scheduling software that constantly changes his or her hours."[22]

It's a powerful narrative that isn't altogether new. In 1981, economist Sherwin Rosen published the paper "The Economics of Superstars," which argued that technological disruptions gave disproportionate power to a few players in any given market. Television, for example, made it possible for the world's highest paid athletes and pop stars to earn exponentially more than others in their fields. Rosen predicted that the rise of superstars would be bad for the bottom line of everyone else, and he was all too right.[23] Today, labor's share of the pie is at its lowest point in half a century. But Silicon Valley companies like Uber, Google, Apple, Facebook, and Amazon—as well as their top-tier workers—are enjoying the superstar effect in spades.[24]

This divide has a massive and underexplored impact not just on individual gig workers, but on the economy at large. Many economists believe that one of the reasons that wage growth remains relatively flat is because of job-disrupting technology itself. Rob Kaplan, the head of the Dallas Fed, believes that technology—and in particular its penetration more broadly and deeply into non-tech industries—is a key reason that we haven't seen wages rising, despite unemployment being at nearly pre–financial crisis lows. What's more, he believes that the Trump corporate tax cuts only exacerbated the trend, as companies incentivized to spend capital on long-term investments put that capital into technology, not people.

"I do about thirty to thirty-five CEO calls with people in and out of the tech sector each month, and it's all about how non-tech firms are implementing technology [in the place of people]." Kaplan believes that we are going to see call centers, airline baggage handlers, reservations agents, and even car dealers replaced by technology in the near future.[25]

The numbers are proving him right. Back in 1998, toward the end of the previous economic expansion, 48.3 percent of business investment went to new structures and industrial equipment (things like factories, machinery, and other brick-and-mortar infrastructure), and about 30 percent went into technology, such as information processing equipment and various types of intellectual property, according to data compiled by Daniel Alpert of Westwood Capital. In 2018, only 28.6 percent of all new investment went to structures and industrial equipment, while technology and intellectual property made up 52 percent.

The difference highlights the shift away from physical investments and toward intangible ones—a trend we see not just in the United States, but also in other wealthy countries like the United Kingdom and Sweden, where investment into intangible assets now exceeds that of tangible ones. The problem is that new factories and machinery tend to create jobs, whereas investments in data processing equipment and software upgrades, which make up a big chunk of current tech-related spending, tend to be job killing, at least in the short term. That can change once workers are able to use the technology to increase their own productivity, as we've already learned. But such an outcome is possible only if education and skill levels keep ahead of the pace of technological change. Sadly, in the United States, education is falling woefully behind the digital revolution.[26]

There are a few sectors, such as finance and information technology, which have seen wages grow. Yet they create relatively few jobs. Finance, for example, takes 25 percent of all corporate profits while creating only 4 percent of jobs. And while half of all American businesses that generate profits of 25 percent or more are tech companies, the tech giants of today—Facebook, Google, Amazon—create far fewer jobs than the big industrial groups of the past, like General Motors and General Electric, but also less than even the previous generation of tech companies such as IBM and Microsoft.

Then there is the growing fear of white-collar job destruction at the hands of Big Tech. A recent study of global executives found that the majority believed they would be retraining or laying off two-thirds of the workforces in the future thanks to digital disruption.

"I think the global professional middle class is about to be blind-sided," says Vivienne Ming, the AI expert I interviewed on the topic in 2018. Ming cites a recent competition at Columbia University between human lawyers and their artificial counterparts to see which group could spot the most loopholes in a series of nondisclosure agreements. "The AI found ninety-five percent of them, and the humans eighty-eight percent," she says. "But the human took ninety minutes to read them. The AI took twenty-two seconds." Game, set, and match to the robots. All of this is one reason why Ming is working with firms such as Accenture to figure out how they can retrain staff to do more creative jobs—the kind that incorporate human emotional intelligence with machine IQ—so that they won't have to lay off hundreds of thousands of accountants, back-office sales staff, and even lower-level programmers in the future.[27]

MEANWHILE, THE EVER-WIDENING gap between the winners and losers is reflected in employee pay as well. Consider that the most profitable 10 percent of U.S. businesses are eight times more profitable than the average company. (In the 1990s, that multiple was just three.) Workers in those super-profitable businesses are paid extremely well, but their competitors cannot offer anywhere near the same packages. Research from the Bonn-based Institute of Labor Economics shows that pay differences between—not within—companies are a major factor in the disparity in worker pay. Another piece of research, from the Centre for Economic Performance in London, shows that this pay differential between top-tier companies and everyone else is responsible for the vast majority of inequality in the United States.

Unsurprisingly, these top sectors and top businesses that take so much of the economic pie tend to be the ones that are the most digitized. As the McKinsey Global Institute's analysis of the haves and have-mores in digital America shows, industries that adopt more technology quickly are more profitable. Tech and finance sit at the top of that chart, whereas sectors that actually create the most jobs—such as retail, education, and government—remain woefully behind. That means you end up with a two-tiered economy: a top level that's very productive, takes the majority of wealth, and creates few jobs, and a bottom one that stagnates.[28]

There are large digital divides along geographic lines as well, which further exacerbates the winner-takes-all trend. For companies to exploit a more entrepreneurial, digitalized economy—whatever sector they operate in—they need access to high-speed broadband, which is three times as likely in urban areas compared with rural ones. There are even big gaps in individual cities. In New York, for example, 80 percent of residents in affluent Manhattan have access to broadband, while only 65 percent of the poorer borough of the Bronx do.[29] The result is a concentration of superstar companies—creating new jobs for superstar workers—in a handful of highly connected cities. Indeed, a 2016 report by the Economic Innovation Group revealed that seventy-five of America's three-thousand-plus counties make up 50 percent of all new job growth. It's a trend that snowballs, as the most talented job seekers are attracted to a handful of cities, driving up property prices and making it tougher for anyone who isn't part of the superstar club to get in the door. This, of course, aggravates the rich-poor divide that is at the heart of partisan politics of the United States, and any number of other countries.[30]

To understand the impact of all this, one has only to visit tech hotbeds like San Francisco or Seattle (or, overseas, places like Tel Aviv, Israel, or Shenzhen, China) and see not only spiraling home prices but the equally spiraling problems of homelessness. The one thing you *won't* see, however, is average middle-class Americans, given that the basics of a middle-class life—housing, healthcare, and retirement savings—are no longer affordable on a middle-class income, thanks to the hordes of paper millionaires created by the tech firms, who increasingly run roughshod over local governments themselves. In Seattle, for example, the city council had proposed imposing a modest $500 per employee tax on local businesses to help address the city's growing homelessness epidemic. But businesses like Starbucks and Amazon complained, and the tax was promptly dropped to $275.[31] And in San Francisco, tech billionaires Jack Dorsey of Twitter, Stripe cofounder Patrick Collison, and Zynga founder Mark Pincus fought tooth and nail against a 2018 ballot proposition to tax companies with revenues of over $50 million a mere 0.5 percent in order to fund local housing and homelessness services. (The measure passed, and was subsequently challenged in court.)

This was brought into focus in 2018 by Amazon's well-publicized search for a second headquarters, which it had to undertake since its own growth was making further expansion in Seattle impossible: Even the company's own workers were finding the inflated prices and unrelenting traffic unbearable. The initial winners were New York and Washington, D.C., which Amazon claimed to have picked on the basis of metrics like the quality of infrastructure, human capital, and transport, even though it rejected many cities that scored well if not better in some such areas. The short list was heavy on locations represented by high-ranking U.S. senators and bids that included billions of dollars in tax credits and other subsidies (both of which NYC and D.C. offered).

Politicians sold the deal, which was contentious in both places, to constituents based on the narrative that Amazon is a huge job creator. But New Yorkers weren't buying it, and not without reason; research shows that communities that offer subsidies to lure big headquarters may see positive headlines and short-term gains, but the end result from an economic standpoint is almost always negative. One recent study found that 70 percent of such city subsidies fall into the category of property tax breaks and job creation tax credits, which means that the big companies pay less for their real estate, but human capital is undermined, because property taxes fund schools and various public services. In other words, the employers that demand skilled workers and good infrastructure are degrading the tax base that creates them. Yet such subsidies have tripled since the 1990s, which leaves states less prepared for economic downturns than they have been in years, thanks to growing municipal debt. Ultimately, it was public outrage over the amount of subsidies being paid by New York City to Amazon that killed the deal; following a spate of public protests, Jeff Bezos decided to pull the HQ offer and leave the city.

Amazon already has vastly more market data than the public sector or any competing retail company, and now, thanks to the sealed bids and nondisclosure agreements that it required officials to sign as part of the HQ2 process, it also has a huge proprietary body of information about the competing cities that it can leverage in whatever way it likes, to whatever financial advance.

Workers Strike Back

Amazon has a big stick. But the one wielded by populists in years ahead may be bigger. Uber has already felt the sharp end of that stick. Its expansion sparked violent protests in Mexico City and Paris, but Kalanick, known for both his detailed grasp of regulatory barriers and the zeal with which he's willing to take them on, was typically unfazed by all of it. "There are a lot of rules in cities that were designed to protect a particular incumbent, but not to move a city's constituents, a city's citizens, and the city itself, forward. And that's a problem," he told me in 2015. "We need to figure out how to merge political progress with actual progress." Uber's idea of progress is simple and sweeping: transportation as ubiquitous and reliable as running water, everywhere, for everyone.

Rewriting the rules for how cities operate, by running roughshod over them if necessary, is, of course, part of that vision. "Who gave the government the right to create monetary value because of scarcity?" said Eric Schmidt to me back in 2015 when I interviewed him for the Kalanick profile. (Google Ventures had invested a whopping $258 million in Uber in 2013, pretty much giving Kalanick a blank check for whatever terms he wanted.)[32] "Cab drivers can't afford million-dollar medallions, so they end up working for financing companies." It's a fair point; while Uber and Lyft have taken much of the blame for the disruption in the taxi industry, recent *New York Times* reporting has also shown that city officials themselves have for years been in cahoots with dicey lenders to drive up the prices of official Yellow Cab medallions, which have since crashed, leaving many drivers in the lurch.[33] Schmidt told me four years ago that he believed people like Kalanick were necessary to disrupt the system. "He fights from a weak position against industrial structures. He is the sort of person that can take nothing and make it into something. People like him are disagreeable in the sense that they disagree [with the status quo]."[34]

Certainly, that's true. Four years ago, I watched Kalanick give a speech to local business leaders in Boston, during which he boldly proclaimed, "I see a world in which there is no more traffic in Boston in five years." Uber's current CEO, Khosrowshahi, has continued to laud the potential traffic- and pollution-reducing effects of the company's reign in urban areas. Yet today, there is disturbing research that shows that ridesharing, while possibly reducing car ownership, may also increase the number of miles traveled in cities, thus having the effect of increasing transit and pollution issues.[35]

Such issues were starting to be apparent when I began covering Uber years ago. Yet Kalanick himself, in typical move-fast-and-break-things mode, wasn't interested in discussing them. In fact, he wasn't terribly comfortable when any such topics were brought up. Like many Silicon Valley types I've known, he would quickly flip into fight mode if you tried to draw him out on some contentious part of the Big Tech debate. His body language would shift and his eyes narrow when I asked about critiques about him and his company, even within the Valley.

"Those people don't know me," he said. "What drives me is a hard problem that hasn't been solved, that has a really interesting and impactful solution. And for me it doesn't even matter what the problem is. I just gravitate toward it. Maybe that results in a style that's a little different," he added, a bit reluctantly. "I'm learning how to be as passionate as I am but understand that when you get bigger, you have to listen more and be more welcoming. And step on toes more lightly." In a particularly revealing moment, he told me he sometimes felt like he was "driving in the fog. I've got my hands on the wheel and I'm going too fast to look behind me, but I can't see very far in front, either."

It was an apt metaphor, and one that could apply to Big Tech as a whole. After all, Uber alone didn't cause the seismic economic shifts roiling the lives of workers everywhere. But, for better or worse, it has benefited from them. Ironically, the shifts wrought by the gig economy are having a positive effect on another group: the labor movement. The labor share of the overall economic pie is at a post–World War II low, which is an enormous problem in an economy that is 70 percent dependent on consumer spending. The demise of the traditional union, where dues are collected by law and workers tend to be in public service or blue-collar areas like building trades and manufacturing, is one of the biggest contributors to that problem. With unions representing only 10.7 percent of the American workforce—half of what it was in the early 1980s—labor simply has no bargaining power these days, an issue exacerbated by the gig economy and automation.

Yet there are signs that a new kind of labor movement may be brewing, one that is broader based, more flexible, and more digital. New York recently launched a $2 million fund to help develop digital cooperative companies such as print shops, neighborhood cafés, and artisan makers of high-end goods. The Freelancers Union, which caters to higher level service workers (writers, graphic artists, photographers) who are being disrupted by the gig economy, now represents some 375,000 workers and is helping to offset declines in traditional unions.

This underscores the changing nature of what it means to be "working class" in the world wrought by Big Tech. If you consider it merely in terms of a dollar per hour figure, many white-collar freelancers would not count. But if you consider it in terms of job security and benefits, as many left-wing economists increasingly do, then these workers absolutely have similar challenges and concerns, from a lack of pensions and health insurance to an increasing vulnerability to being undercut by job-replacing technologies, which are moving higher up the economic food chain.

The economic and political potential of that mix appeals to Freelancers Union founder Sara Horowitz. "Ideologically, I come out of the Jewish labor movement of the 1920s, which included not only garment factory workers but also small entrepreneurs," she says. Indeed, she has brought an interesting mix of entrepreneurial zeal and strategic thinking to her community. The Freelancers Union was, for example, instrumental in getting a law passed in New York City that allows independent contractors to sue for double damages and legal fees when clients fail to pay. Horowitz's group subsequently developed an app to help members find lawyers who would take on their cases, and since most of those legal professionals were independent or working for small firms, she began organizing them, too. (She plans to take this strategy into other areas, including the accounting profession.) "I get concerned when the labor movement is defined narrowly," says Horowitz, who wants Democrats to better address the shared concerns of a full range of workers of all incomes.

It could be an opportune moment to do so. A Pew Foundation study shows that millennials have different views about unions than their parents do. Favorable feelings about unions have been growing steadily since they hit their low point around 2010; today, 48 percent of the population believes that unions are a good thing, and millennials have the most positive view. As Kashana Cauley, a writer for *The Daily Show with Trevor Noah,* wrote in *The New York Times,* millennials are heeding the call to fight back at a time when "the government wants to disembowel public and private healthcare and when wages are on the decline," by joining "existing unions or unionizing ourselves." Interestingly, this view has traction among young people on both sides of the political aisle. Half of conservative millennials actually support organized labor, compared with only 24 percent of older Republicans. This presents a major opportunity for Democrats in 2020 if they can keep their party from being captured by the coastal libertarians of Silicon Valley or Wall Street.[36]

When I last spoke with Kalanick, he'd gotten interested in another period of labor disruption: the late 1800s, better known as the Gilded Age. Appropriately, he told me he was currently reading biographer Ron Chernow's book on John D. Rockefeller, *Titan*. Rockefeller was, like Kalanick, a self-made man who eventually created the world's largest and most powerful monopoly, Standard Oil, beating down regulators, unions, and political officials in the process.[37] It's a story that many politicians and regulators are now looking to, as they grapple with the new monopolists of Big Tech.

CHAPTER 9

The New Monopolists

After my brief flirtation with the idea of working at Google nearly a decade ago, the next time I'd visit the company's New York office was in 2017, shortly after taking up my position as the *Financial Times'* global business columnist. The food was just as good, but the cognitive dissonance between how the Googlers viewed the company and how many others viewed it had grown more extreme. When I brought up the issue of monopoly power with the public policy staffer I was meeting with, she seemed genuinely surprised. "We feel like we are under threat all the time, from other big technology companies," she told me. "We just don't understand the argument that there's not enough competition in the marketplace."

I could see it from her perspective—as we've already learned, Amazon has become a major competitor for Google. The problem is that two or three behemoths competing against each other doesn't really count as a competitive economic landscape. For that, you need a wide variety of firms of all sizes able to enter and thrive in a marketplace. But increasingly, that's not happening. That's because Big Tech enjoys several natural advantages that breed monopoly power: information asymmetries, the network effect, the ability to easily copy the ideas of smaller competitors in an open-source environment, gatekeeper rents (even if they are in the form of data, not dollars), the advantages of both owning a platform that others must use as well as being able to conduct commerce on that platform, and the legal and political muscle the largest players can exert at will in Washington. I would witness that political power firsthand later that year, as I watched Eric Schmidt, one of the largest funders of the New America Foundation, an influential Washington think tank, squash a policy wonk whose ideas he found threatening.

I had known of the scholar Barry Lynn because of his prescient work on supply chain economics, in which he'd examined how the United States had lost manufacturing competitiveness to China.[1] Lynn, who like me was an advocate for localism and small business, had recently begun looking at the way in which Big Tech firms were dominating the economy, and hindering entrepreneurial vibrancy and growth as a result. When Lynn's research group, the Open Markets division, posted an article on the think tank's website in praise of the EU antitrust ruling against Google, Schmidt (who was at the time still executive chairman of Alphabet, Google's parent company) called up the head of New America, Anne-Marie Slaughter, a former director of policy at the State Department under Hillary Clinton, and voiced his disapproval.

That was when Slaughter told Lynn that "the time has come for Open Markets and New America" to part ways. Not because of his work, she stressed in an email that was reviewed by *The New York Times*, but because his "lack of collegiality" was imperiling the organization as a whole. The episode called to mind an email that Slaughter had sent him a year earlier, in 2016, in the run-up to a well-received conference on the market dominance of Google, Amazon, and Facebook, which Lynn had organized. Apparently, Slaughter indicated, Google was concerned that its position wouldn't be represented. "We are in the process of trying to expand our relationship with Google on some absolutely key points," Slaughter wrote to Lynn. She urged him to "just THINK about how you are imperiling funding for others."[2]

Lynn was eventually pushed out of the think tank (both Google and Slaughter deny this was due to Schmidt's pressure)[3] and ended up starting what has become an even more influential stand-alone think tank, the Open Markets Institute (full disclosure: I sit on the advisory board of the new entity). And his concerns about Big Tech's monopoly power have come to the forefront of the policy conversation in Washington, influencing liberals and conservatives alike. One of the most influential pieces of work associated with the think tank was by a young legal scholar named Lina Khan, whose paper, "Amazon's Antitrust Paradox," published in January 2017 by the *Yale Law Journal,* laid out why the old ways of thinking about monopoly power are no longer adequate for the digital era.[4] (Khan worked with Open Markets for two years before law school and a year afterward as well.)

It's hard to believe that Khan, an unassuming thirty-year-old scholar, working in the long-neglected field of antitrust law, is currently public enemy number one for the world's tech titans (or perhaps number two, just behind EU competition commissioner Margrethe Vestager, who has taken some of Khan's ideas on board). Khan's breakthrough came from her "interest in the economists who were actually studying power....That's something that gets channeled out in the contemporary version of economics," she said during an interview with me in 2019.

Academic papers don't tend to go viral, but this one received a near-unprecedented level of interest from policy makers. In fewer than one hundred pages, Khan made the case that current interpretations of U.S. antitrust law, which is meant to regulate competition and curb monopolistic practices, are utterly unsuited to the architecture of the modern economy.

For roughly four decades, antitrust scholars—taking their lead from Robert Bork's 1978 book *The Antitrust Paradox*—have pegged their definitions of monopoly power to short-term price effects; so if Amazon is making prices lower for consumers, the market must be working effectively. Khan set out a simple but powerful counter-argument: that it doesn't matter if companies such as Amazon are making things cheaper in dollars if they are using predatory pricing strategies to dominate multiple industries and choke off competition and choice. "I was fascinated by Wall Street's view of Amazon, and how much it differed from what conventional economic theory would say about the company," Khan said.

Her reframing of the problem was revelatory, and is now at the core of a number of Big Tech antitrust actions on both sides of the Atlantic. Experts have been telling us for years that the concentration of power in a few corporate giants was a crucial factor in everything from wage stagnation to rising inequality and political populism.[5] Now, suddenly, with Khan's paper, there was a road map for understanding the problem, and a potential legal tool for dealing with it, too.

"At a basic level, I'm interested in imbalances in market power and how they manifest. That's something you can see not just in tech but across many industries," said Khan, who has written sharp pieces on monopoly power in areas as diverse as airlines and agriculture. "A lot of people talk about markets as these forces that are the product of globalization and technology and these things that are totally unrooted, that are totally separate from laws and legal institutions." Khan, like many in her cohort, believes otherwise. "If markets are leading us in directions that we, as a democratic society, decide are not compatible with our vision of liberty or democracy, it is incumbent upon government to do something."[6]

The paper focused, as the title indicates, on Amazon, which is by many measures the most powerful and dominant of the FAANGs. Today, Amazon controls the largest single chunk of e-commerce in the United States, and lives up to the title of Brad Stone's book: *The Everything Store*. But as many will recall, the company began as a mere purveyor of below-cost books, one that threatened to bring down publishers—not to mention brick-and-mortar bookstores—with pricing techniques that could only be described as predatory. The tactics Amazon employed in its full-frontal assault on the book business would soon become standard operating procedure as the company went on to launch similar offensives in countless other industries and markets.

The Illusion of "Cheap"

If there's one thing Amazon's approach in taking on the book business wasn't, that word would probably be *subtle*. As Stone laid out in his book,[7] Bezos directed employees to "approach these small publishers the way a cheetah would a sickly gazelle."[8] The Gazelle Project, as it was named, involved deeply discounting bestsellers to establish dominance in the ebook market, in a way that is similar to how Apple came to dominate digital music. Amazon also sold its Kindle reading device below cost; both techniques helped to build a network of consumers that would then be drawn back time and time again to the e-tailer. At the same time, the company would discount print books sold online, too, using exactly the sort of predatory pricing techniques that Khan lays out in her paper. The company may have been earning razor-thin margins on each ebook sale, but the strategy worked. By 2009, roughly two years after the Kindle launched, Amazon was selling around 90 percent of all ebooks.

The major publishers feared that Amazon's pricing strategy was changing consumers' perceptions about what was a "normal" price to pay for an ebook in a way that would permanently alter the economics of the book business, and tried to wrest back some control. Five of the "Big Six," as the largest publishers were known before Penguin merged with Random House in 2013, decided to try to shift business to Apple, which agreed to let publishers set the price consumers would be charged—meaning Apple couldn't slash the price of an ebook in half whenever it wanted—and take a 30 percent cut from whatever price the publishers agreed was fair. Macmillan, one of the Big Six, pushed Amazon to accept this deal, too, at which point Amazon turned around and accused Macmillan of exerting monopoly power.

The irony of a book publisher worth a fraction of what Amazon is worth being able to exert any kind of relative power in the market was lost on the Department of Justice, which decided to sue both Apple and the publishers in 2012 in an antitrust suit that claimed collusion. Various outside experts and politicians complained that the DOJ was going after the wrong party—after all, what were Amazon's pricing strategies if not an attempt to capture more and more market share?—but investigators found "persuasive evidence lacking" that the company had engaged in predatory pricing, since its business had been consistently profitable, even when books were steeply discounted.

The problem, according to Khan, was that this approach didn't take into account two things. First, discounting products sold on a digital platform like Amazon affords that platform owner certain advantages that a retailer discounting products in traditional stores doesn't enjoy—namely the data garnered by those ultralow prices (which Amazon gathered even when customers were simply browsing on its platform and not even buying). And second, the multiple ways in which Amazon—which by that time had grown to dominate a number of other retail sectors—could recoup the losses it willingly took on ebooks. The DOJ was looking at pricing power in an extremely linear way: Was a particular business line (books, say, or diapers, or white goods) losing money in order to undercut the competition, and were consumers suffering as a result?

But platform technologies had changed the publishing business— and indeed, every retail business—in ways that made the old ideas of pricing irrelevant. "What the DOJ missed," explained Khan, "is the way in which below-cost pricing . . . entrenched the reinforced Amazon's dominance in ways that loss leading by physical retailers does not."

Buoyed by the success of this strategy, Amazon has since used similar tactics to dominate so many other areas, ultimately undercutting competitors not just in traditional retail, but in e-commerce as well. In the market in baby products, for example, Amazon drove a competitor called Quidsi out of the dominant position by using bots to monitor Quidsi's prices—and then knocking its own down by the optimal amount, in real time. Amazon eventually bought up the company, just as it has many competitors, like shoe retailer Zappos, wholesale.

Amazon is now the default starting point for online shopping, accounting for 44 percent of U.S. consumers' first search for products according to one study.[9] In addition to being a retailer, it is also a marketing platform, a delivery and logistics network, a payment service, a credit lender, an auction house, a major book publisher, a producer of television and films, a fashion designer, a hardware manufacturer, and a leading host of cloud server space and computing power.[10] It has racked up double-digit increases in net sales yearly, for several years running, all the while accepting operating losses or low margins in order to attain even more market share, in ever more industries.

Amazon's prices are seductive—truth be told, I shop there frequently, and I expect many of the people reading this book do, too. But it's not up to consumers to enforce the competitive landscape; it's up to regulators. And as they look more closely at Amazon, there is plenty of behavior that looks anticompetitive, if not downright creepy. Consider, for example, the way in which Alexa can direct us to certain products and away from others. One study found that such nudges could create a 29 percent increase in sales for Amazon.[11]

But remember that prices aren't really so cheap on Amazon if you consider the value of the data you are giving up. One conservative estimate of the value of personal data to platform firms like Google, Facebook, and Amazon, as well as to other big aggregators like credit bureaus, put the number at around $76 billion in 2018.[12] And that's just tabulating the way in which it allows such firms to sell targeted advertising (which is half of the overall ad revenue for platforms).[13] Those figures don't count the way in which all the personal data can be collated to increase its value, or employed by the firms in a variety of ways to nudge you toward certain purchasing decisions.

A close read of Google chief economist Hal Varian's 1998 book, *Information Rules,* makes it clear that the people in charge of such firms knew that "free" products were actually an illusion. The problem is that we have no clear idea how valuable our data actually is to the companies that mine it. "Why does Google give away products like its browser, its apps, and the Android operating system for mobile phones?" asked Varian rhetorically in a 2009 *Wired* piece. "Anything that increases Internet use ultimately enriches Google."[14] By "giving" things away in exchange for something that is actually much more valuable, companies like Google and Amazon bring in enormous profits, while also creating impenetrable moats around their businesses.

Amazon now controls so much of the logistics industry that it can demand steep discounts—as much as 70 percent off the going rates—from companies like UPS. That leaves delivery companies to charge other, smaller customers more in order to make up for the cut in margins.[15] In yet another amazing competitive jujitsu move, Amazon has started a new business offering logistics and delivery services to the very retailers who are now being charged higher prices by UPS and FedEx as a result of Amazon. Merchants who sign up, most of whom are already competing with Amazon itself for sales, now find themselves at even less of a competitive advantage, thus further strengthening Amazon in the process.

Amazon is like the house in a Vegas casino—it always wins. "You can't really be a high-volume seller online without being on Amazon, but sellers are very aware of the fact that Amazon is also their primary competition," complained one merchant to *The Wall Street Journal* in 2015.[16] The end result is Amazon controlling a bigger piece of more and more markets. Bezos's retail behemoth is now a fulfillment and logistics juggernaut, with thousands of trucks, container ships, planes, and drones at its disposal. Former employees have said that its ultimate goal is to supplant all delivery services, making it not just the Everything Store, but the Everything Shipper.[17]

The rise of tech platforms has been linked to a decline in new business growth and reduced opportunities for entrepreneurs.[18] This is in part because platforms can move quickly into new business lines in a way that traditional businesses, particularly smaller ones, cannot. Remember from chapter 7 the way in which Google was able to copy Yelp's local search business model and quickly move to take the space (and the subsequent ad revenue) for itself? Again, like the railroads or telecom companies of the past, Big Tech is able to both create a market and conduct commerce within it—and that is clearly an unfair advantage.

Technology firms are widely understood to enjoy unusual and (many would say) unfair profits because of their monopoly power. In 2018, *The Economist* calculated that increased corporate concentration has led to a pool of $660 million in abnormal profits, two-thirds of which came from the United States, and one-third of that from the tech sector alone.[19] This is a direct result of the way in which they can upend the usual economic laws of gravity.

Big Tech makes a big point of denying this. "We are one click away from losing you as a customer, so it is very difficult for us to lock you in as a customer in a way that traditional companies have," said Google's Eric Schmidt back in 2009.[20] Yet a spate of research shows that customers are unlikely to switch platforms once they've reached a dominant position in the marketplace, because "switching costs" are quite high—it's a cognitive pain to move from one platform to another, in a way that is different from simply deciding to do one's shopping at a different store. (Just think about the trouble of password memorization alone.)[21] The truth of the matter is that most of us would be more likely to go out and take a walk around the block than we would be to switch to, say, Microsoft's Bing search engine if Google suddenly went down. The usual laws of competition simply don't work in the world of platforms.

The Antitrust Paradox

When a company becomes a player in a market while also owning the marketplace itself, it's clearly problematic from a competitiveness standpoint. This is why there are rules in the financial sector to prevent companies from owning assets they are trading, and from trading in markets they created (though these rules are sometimes circumvented by clever lawyers and lobbyists). Technology companies have so far avoided such specialized restrictions, even as they've become some of the largest and most concentrated companies in the world.[22] This is due in part because the opacity of their business model makes them hard to even understand, let alone regulate. But it's also because regulators who might curb their power are working with an outdated model of monopoly power, one that hasn't been revisited in forty years.

Indeed, the last time we had a major reset of antitrust policy in the United States was when Robert Bork published *The Antitrust Paradox* in 1978. Bork held that the major goal of antitrust policy should be to promote "business efficiency," which from the 1980s onward came to be measured in consumer prices. It was a shift that took the United States away from antitrust policy predicated on the welfare of the "citizen" and toward one that clearly served the laissez-faire politics of the Reagan administration. The problem is that in a world where data is the new currency, price is an insufficient—if not irrelevant—metric. This has provoked calls for an overhaul of antitrust policy similar to the one America had with the passage of the Sherman Act at the end of the nineteenth century, which was designed to ensure that the economic power of large companies did not result in the corruption of the political process.

It's a timely call. Income inequality and corporate consolidation in the United States have reached levels not seen since that Gilded Age, which is no accident, since our monopoly laws have become just as weak and inefficient as they were back then. At the turn of the last century, oligopolies such as Standard Oil and U.S. Steel were in many ways more powerful even than the government. They often had paid politicians in their pockets; President William McKinley "tacitly acknowledged that Wall Street rather than the White House had executive control of the economy,"[23] just as the many players in the tech sector arguably do today.

The robber barons of the Gilded Age were eventually stymied by Louis Brandeis, the advocate, reformer, and Supreme Court justice who grew up around the mid to late 1800s in Louisville, Kentucky, a diverse and decentralized midsized American town that Brandeis praised as "idyllic" and free from the "curse of bigness" (a Brandeis phrase that Columbia University legal scholar Tim Wu, who advocates for a return to turn-of-the-century antitrust interpretations, has repopularized).[24]

The Louisville of Brandeis's youth was prosperous, but relatively untouched by the sort of industrial concentration found on the coasts and some other parts of the country. It was a place where small farmers, retailers, professionals, and manufacturers all knew one another, worked together, and had the sort of shared moral framework that Adam Smith believed was a key to well-functioning markets. But by the time Brandeis himself became a lawyer in Boston, oligarchs John D. Rockefeller and J. P. Morgan were building empires—Rockefeller's oil dynasty and Morgan's railroad monopoly—that were neither moral nor efficient. But the tycoons had bought the legislatures, and there was no one powerful enough to reel them in.

Brandeis boldly took them on, in a case against Morgan's New Haven Railroad that exposed the underside of monopoly power: cartel pricing, bribes to officials, accounting fraud, and so on. The result was not only the breakup of the railways, but a new approach to antitrust, and a public belief in the idea that government should, as Wu puts it, "punish those who used abusive, oppressive, or unconscionable business methods to succeed." Brandeis believed that in limiting individuals' ability to work and compete and prosper on their own terms, giant corporations were robbing people of their very humanity. As he wrote, "Far more serious even than the suppression of competition is the suppression of industrial liberty, indeed of manhood itself."

This philosophy, which was brought into the mainstream by conflicted trust buster Theodore Roosevelt (who both loved and loathed power, but wanted to see corporations curbed by government), lasted through the 1960s. But with the rise of conservative Chicago School academics, in particular Robert Bork, the notion that too much corporate power alone was problematic soon fell out of favor. Antitrust policy became technocratic and weak, pegged to the idea that as long as companies reduced prices for consumers, they could be as big and powerful as they wanted.

That fundamental shift has, of course, allowed any number of industries, from airlines to media to pharmaceuticals, to reach unprecedented levels of consolidation. Yet it is the tech business more than any other—in which products and services aren't just cheap but "free," or rather bartered in exchange for personal data in opaque transactions—that illustrates the need for a new interpretation of monopoly power.

To Wu, Khan, Lynn, and an increasing number of other experts, Google, Facebook, and Amazon are the Standard Oil or U.S. Steel of our day—companies that are more powerful than governments, and ones that pose a threat to liberal democracy unless they can be curbed through a broader view of monopoly. Given the unique challenges Big Tech poses, the new measure of antitrust action should not only include a broader view of consumer pricing and welfare, but whether new companies have the ability to enter a market controlled by the tech monopolists and have their product compete on its merits.

"Much of the time," says Lina Khan, "that answer will be no."[25] Khan is examining a host of old cases—from railroad antitrust suits to the separation of merchant banking and the ownership of commodities—to argue that "if you are a form of infrastructure, then you shouldn't be able to compete with all the businesses dependent on your infrastructure."[26]

New rules can't come soon enough, because the growth of the Big Tech firms has triggered a domino effect in concentration in the *rest* of the economy, something that many economists believe has become a big headwind slowing shared growth. Between 1997 and 2012, corporate concentration rose in two-thirds of the nine hundred or so census-surveyed industries, with the weighted average market share of the top four firms in each industry growing from 26 percent to 32 percent.[27] That's because companies of all stripes believe they need more heft to play against the FAANGs.

Over the past few years, even giants in old guard industries have struggled to maintain the scale they believe is necessary to compete. A record number of corporate mergers and acquisition deals were inked in 2018, many of which involved big companies trying to compete with even bigger digital companies that had disrupted their traditional business models. CVS buying Aetna, for example, was a reaction to Google and Amazon moving into the healthcare space. Walmart's purchase of Flipkart, a major Indian grocer, came after Amazon gobbled up Whole Foods.

Nowhere is this phenomenon more apparent than in media and telecoms.[28] Consider the fight between Disney and Comcast for the assets of 21st Century Fox, or the proposed merger that T-Mobile and Sprint pitched to the Federal Communications Commission in 2018. Perhaps most important, a contentious U.S. District Court decision that same year allowed AT&T and Time Warner to merge, opening the floodgates to a host of new deals.[29] As district court judge Richard J. Leon, who approved the deal, put it in his 172-page opinion: "If there ever were an antitrust case where the parties had a dramatically different assessment of the current state of the relevant market, and a fundamentally different vision of its future development, this is the one. Small wonder it had to go to trial!"

It's tough to argue that two cable giants teaming up is a good thing for consumers. Yet the deal underscores the way in which the media landscape has changed dramatically in the past few years. As hard as it might be to believe, AT&T and Time Warner, which together form a multibillion-dollar media conglomerate, are still small potatoes next to their new Silicon Valley competitors: streaming services Netflix and Amazon, as well as Facebook, Google, and, most recently, Apple, which in 2019 announced a big move into entertainment and media.

Makan Delrahim, the Department of Justice's head of antitrust, argued that AT&T should be prevented from buying Time Warner because the two companies' merger would result in higher cable prices for American consumers. Yet the judge bought the companies' claim that the merger was necessary to stave off competitive pressure from those bigger fish: Google offers up to fifty channels of premium content on YouTube for $49.99 a month. Amazon and Netflix have invested heavily in original content (Netflix put $13 billion toward content development in 2018 alone) to compete for both viewers and talent with HBO. Apple and Facebook each spent $1 billion in 2018 on original video content.

In 2017, Google and Facebook took 84 percent of the digital advertising market. As Judge Leon noted on page two of his Time Warner decision, "Facebook's and Google's dominant digital advertising platforms have surpassed television advertising revenue," making it even tougher for companies like Time Warner to keep subscription fees low. No wonder more than 22 million U.S. cable customers "cut the cord," or got rid of their cable boxes, in that same year—up 33 percent from 2016.[30] If someone has monopoly power in this world of digital media, it is not the legacy media players.

But the tide may be starting to turn, as regulators finally begin to wake up to the competitive threat posed by corporate concentration. In 2017, the European Union, which has taken the lead on monopoly issues, hit Google with a record antitrust fine, $2.7 billion, for unfairly favoring its own services over those of its rivals. The core complainant came from the U.K.-based shopping service

Foundem, discussed in chapter 7, but the case also touched on issues relevant to the Yelp-Google conflict and the FTC case against Google that was dismissed in 2012. Margrethe Vestager, the EU competition commissioner, hit the point home with a strongly worded letter accusing the company of "destroying jobs and stifling innovation." The very next year, the European Union hit Google with an even bigger fine, some $5 billion, for abusing its dominance in the mobile market.[31]

Obviously, antitrust litigation is a slow and complex process. But even in the United States, where the last big antitrust case was taken on more than twenty years ago, there are finally signs of change. Joseph Simons, the chairman of the Federal Trade Commission, has pledged "vigorous" antitrust enforcement, and in 2018 convened hearings on competition and consumer protection, the first broad policy hearings on the topic since 1995. House Democrats in particular are galvanizing around the issue. Even Republican lawmakers joined Democrats in calling on the FTC and the Department of Justice to investigate the largest tech companies. In July of 2019, Facebook disclosed that the FTC had indeed begun an antitrust investigation into the company. Both the FTC and the DOJ are also looking into possible actions involving other Big Tech firms.

Makan Delrahim, the DOJ head of antitrust who tried to prevent the AT&T–Time Warner merger, has told me that he believes that price isn't the only metric for consumer welfare, and that "data is an important asset." While he is not opposed in principle to Big Tech's business models or dealmaking, he is concerned about abuse of a dominant position. Many critics see evidence of that behavior in Google today, something that Delrahim has told me the DOJ is looking out for.[32] "Can you use position to disadvantage and discriminate against a new technology that would challenge a monopoly position?" he asks. "I think that's an important test and a good guidepost for a lot of us as we look to the type of practices that are happening with a Google or with anyone else."

A Price on Data?

The big question now is *how* policy should shift, and on what basis new antitrust and monopoly cases should be argued. There are some who believe the Chicago School's consumer pricing philosophy could actually be used to curb the power of the tech titans. "As data becomes more and more important, you get more efficient products for consumers, but you also get certain barriers [to competition]," says Delrahim. "There should be competition to create and collect data," he says, hinting at the notion that choice—and not just price—should be part of the consumer welfare metric.[33] SEC commissioner Robert Jackson has said he believes that companies should have to report the value of their data on their filings, just as they would any other material holding. If that information were publicly available, it would go some way toward clarifying the true power of Big Tech in the marketplace.

That would, of course, require putting a price on data—and efforts to do just that are now under way. As investor Roger McNamee points out, it's true that the combined power of data and the network effect have created value for consumers. And yet, they've created exponentially more value for Big Tech companies. "Each time Google introduced a new service, consumers got a step function increase in value, but nothing more. Each new search, email message, or map query generates approximately the same value to the user," he says. "Meanwhile, Google receives at least three forms of value: whatever value it can extract from that data point through advertising, the geometric increase in advertising value from combining data sets, and new use cases for user data made possible by combining data sets. One of the most valuable use cases that resulted from combining data sets was anticipation of future purchase intent based on a detailed history of past behavior. When users get ads for things they were just talking about, the key enabler is behavior prediction based on combined data sets."[34]

Bottom line? Consumers are giving up more value in personal data than they receive in services—indeed, vastly more. Which means that the true price of Big Tech to each of us has been rising sharply—in lockstep with the amount of time we spend on our devices and thus the amount of data we've generated over twenty years. And if that can be shown to be the case, then regulators can make a strong argument, even with the current Chicago School thinking, that Google, Facebook, Amazon, and other giants fail to meet the standard of consumer welfare and should be regulated in new ways, or broken up.

But there are many others—and I would put myself among them—that believe we need to go beyond the Chicago School, and consider more deeply the ways in which Big Tech's power has distorted markets and the political economy. In his book *The Curse of Bigness*, academic Tim Wu has made a persuasive case for neo-Brandeisian reforms that would include things like broader public hearings and debate over big mergers, forced breakups of mergers that are ultimately found to be uncompetitive (he advocates for spinning both Instagram and WhatsApp off from Facebook), and new rules that would allow regulators to investigate not just individual firms but entire economic landscapes (a method that was used in the United Kingdom to determine that joint ownership of the Heathrow, Gatwick, and Stansted airports was not serving the public).[35]

He and others, including Open Markets' Barry Lynn, and Lina Khan (who has advised the FTC and joined the staff of the House Judiciary Subcommittee on Antitrust, Commercial and Administrative Law, which has begun hearings on the topic of Big Tech and antitrust), also argue for dropping consumer welfare in favor of "citizen welfare" as the bar for mergers. "Decades of practice have shown that the promised scientific certainty of the Chicago method has not materialized, for economics does not yield answers, but arguments," writes Wu.

Fair point. Indeed, if the 2008 economic crisis didn't put the final nail in the coffin of the Chicago School's own monopoly of ideas in economics, then the rise of the digital giants certainly should. Both have contributed to the feeling among many ordinary Americans that the system is rigged. And that's not good for the economy or for our democracy. "The new Brandeis movement isn't just about anti-trust," says Khan. Rather, it is about values. "Antitrust laws used to reflect one set of values, and then there was a change in values that led us to a very different place." Now that corporate power in this country has reached levels not seen since the Gilded Age, it's time for another such change.

Whether or not Washington will listen remains to be seen. Yet one thing is clear: There is more than just economic vibrancy at stake here. Whether by antitrust policies, by agency regulation, or by some new philosophy of welfare, Silicon Valley's economic and political power should be curbed, lest we fail a very costly stress test of democracy.

CHAPTER 10

Too Fast to Fail

The late, great management guru Peter Drucker once said, "In every major economic downturn in U.S. history the 'villains' have been the 'heroes' during the preceding boom."[1] I can't help but wonder if that might be the case over the next few years, as the United States (and possibly the world) heads toward its next big slowdown. Downturns historically come about once every decade, and it's been more than that since the 2008 financial crisis. Back then, banks were the "too-big-to-fail" institutions responsible for our falling stock portfolios, home prices, and salaries. Technology companies, by contrast, have led the market upswing over the past decade. But this time around, it's the Big Tech firms that could play the spoiler role.

You wouldn't think it could be so when you look at the biggest and richest tech firms today. Take Apple. Warren Buffett says he wished he owned even more Apple stock. (His Berkshire Hathaway has a 5 percent stake in the company.) Goldman Sachs is launching a new credit card with the tech titan, which became the world's first $1 trillion market cap company in 2018. But hidden within these bullish headlines are a number of disturbing economic trends of which Apple is already an exemplar. Study this one company, and you begin to understand how Big Tech companies—the new too-big-to-fail institutions—could indeed sow the seeds of the next crisis.

THE FIRST THING to consider is the financial engineering done by such firms. Like most of the largest and most profitable multinational companies, Apple has loads of cash—$285 billion—as well as plenty of debt (close to $122 billion). That is because—like nearly every other large, rich company—it has parked most of its spare cash in offshore bond portfolios over the past ten years. At the same time, since the 2008 financial crisis it has issued debt at cheap rates to do record amounts of share buybacks and dividend payments. Apple is responsible for about a quarter of the $407 billion in buybacks announced since the Trump tax bill was passed in December 2017.[2]

But buybacks have bolstered mainly the top 10 percent of the U.S. population that owns 84 percent of all stock.[3] The fact that share buybacks have become the single largest use of corporate cash for over a decade now has buoyed markets. But it has also increased the wealth divide, which many economists believe is not only the biggest factor in slower-than-historic trend growth, but is also driving the political populism that threatens the market system itself.

That phenomenon has been put on steroids by yet another trend epitomized by Apple: the rise of intangibles such as intellectual property and brands (both of which the company has in spades) relative to tangible goods as a share of the global economy. As Jonathan Haskel and Stian Westlake show in *Capitalism Without Capital,* this shift became noticeable around 2000, but really took off after the introduction of the iPhone in 2007. The digital economy has a tendency to create superstars, since software and Internet services are so scalable and enjoy network effects. But according to Haskel and Westlake, it also seems to reduce investment across the economy as a whole. This is not only because banks are reluctant to lend to businesses whose intangible assets may simply disappear if they go belly-up, but also because of the winner-takes-all effect that a handful of companies, including Apple (and Amazon and Google), enjoy.

As we read in the last chapter, that's likely a key reason for the dearth of start-ups, declining job creation, falling demand, and other disturbing trends in our bifurcated economy. Concentration of power of the sort that Apple and Amazon enjoy is a key reason for record levels of mergers and acquisitions. In telecoms and media especially, many companies have taken on significant amounts of debt in order to bulk up and compete in this new environment of streaming video and digital media.

Some of that high yield debt is now looking shaky, which underscores that the next big crisis probably won't emanate from banks, but from the corporate sector. Rapid growth in debt levels is historically the best predictor of a crisis. And for the past several years, the corporate bond market has been on a tear, with companies in advanced economies issuing a record amount of debt; the market grew 70 percent over the last decade, to reach $10.17 trillion in 2018.[4] Even mediocre companies have benefited from easy money. But as the interest rate environment changes, perhaps more quickly than was anticipated, many could be vulnerable. The Bank for International Settlements—the international body that monitors the global financial system—has warned that the long period of low rates has cooked up a larger than usual number of "zombie" companies, which will not have enough profits to make their debt payments if interest rates rise. When rates eventually do rise, losses and ripple effects may be more severe than usual, warns the BIS.

Of course, if and when the next crisis is upon us, the deflationary power of technology exemplified by companies like Apple could make it more difficult to manage. That is the final trend worth considering. Technology firms drive down the prices of lots of things, and tech-related deflation is a big part of what has kept interest rates so low for so long; it has not only constrained prices, but wages, too. The fact that interest rates are so low, in part thanks to that tech-driven deflation, means that central bankers will have much less room to navigate through any upcoming crisis. Apple and the other purveyors of intangibles have benefited more than other companies from this environment of low rates, cheap debt, and high stock prices over the past ten years. But their power has also sowed the seeds of what could be the next big swing in the markets.[5]

The New Too-Big-to-Fail Firms

All of this reminds me of a fascinating conversation I had a few years ago with an economist at the U.S. Treasury's Office of Financial Research, a small but important body that was created following the 2008 financial crisis to study market trouble, and which has since seen its funding slashed by President Trump. I was trawling for information about financial risk and where it might be held, and the economist told me to look at the debt offerings and corporate bond purchases being made by the largest, richest corporations in the world, such as Apple or Google, whose market value now dwarfed that of the biggest banks and investment firms.[6]

In a low interest rate environment, with billions of dollars in yearly earnings, these high-grade firms were issuing their own cheap debt and using it to buy up the higher-yielding corporate debt of other firms. In the search for both higher returns and for something to do with all their money, they were, in a way, acting like banks, taking large anchor positions in new corporate debt offerings and essentially underwriting them the way that J.P. Morgan or Goldman Sachs might. But—it's worth noting—since such companies are not regulated like banks, it is difficult to track exactly what they are buying, how much they are buying, and what the market implications might be. There simply isn't a paper trail the way there is in finance. Still, the idea that cash-rich tech companies might be the new systemically important institutions was compelling.

I began digging for more on the topic, and about two years later, in 2018, I came across a stunning Credit Suisse report that both confirmed and quantified the idea. The economist who wrote it, Zoltan Pozsar, forensically analyzed the $1 trillion in corporate savings parked in offshore accounts, mostly by Big Tech firms. The largest and most intellectual-property-rich 10 percent of companies— Apple, Microsoft, Cisco, Oracle, and Alphabet among them— controlled 80 percent of this hoard.[7]

According to Pozsar's calculations, most of that money was held not in cash but in bonds—half of it in corporate bonds. The much-lauded overseas "cash" pile held by the richest American companies, a treasure that Republicans under Trump had cited as the key reason they passed their ill-advised tax "reform" plan, was actually a giant bond portfolio. And it was owned not by banks or mutual funds, which typically have such large financial holdings, but by the world's biggest technology firms. In addition to being the most profitable and least regulated industry on the planet, the Silicon Valley giants had also become systemically crucial within the marketplace, holding assets that—if sold or downgraded—could topple the markets themselves. Hiding in plain sight was an amazing new discovery: Big Tech, not big banks, was the new too-big-to-fail industry.

Growth over Governance

As I began to think about the comparison, I found more and more parallels. Some of them were attitudinal. It was fascinating, for example, to see how much the technology industry's response to the 2016 election crisis mirrored the banking industry's behavior in the wake of the financial crisis of 2008. Just as Wall Street had obfuscated as much as possible about what it was doing before and after the crisis, every bit of useful information about election meddling had to be clawed away from the titans of Big Tech.

First, they insisted that they'd done nothing wrong, and that anyone who thought they had simply didn't understand the technology industry. This completely mirrors the "You just don't get it" attitude that critics of the financial sector face. It was under extreme pressure from both press and regulators that Facebook's Mark Zuckerberg finally turned over three thousand Russia-linked ads to Congress. Google and others were only marginally less evasive. Similar to Wall Street financiers at the time of the U.S. subprime crisis, the tech titans have remained, years after the 2016 election, in a largely reactive posture, parting with as few details as possible, attempting to keep the asymmetric information advantages of their business model that, as in the banking industry, help generate outsized profit margins. It's a "deny and deflect" attitude that is similar to what we saw from financiers in 2008, and has resulted in deservedly terrible PR.

But there are more substantive similarities as well. At a meta level, I see four major likenesses in Big Finance and Big Tech: corporate mythology, opacity, complexity, and size. In terms of mythology, Wall Street before 2008 sold us on the idea that what was good for the financial sector was good for the economy. Until quite recently, Big Tech tried to convince us of the same. But there are two sides to the story, and neither industry is quick to acknowledge or take responsibility for the downsides of "innovation."

A raft of research shows us that trust in liberal democracy, government, media, and nongovernmental organizations declines as social media usage rises.[8] In Myanmar, Facebook has been leveraged to support genocide. In China, Apple and Google have bowed to government demands for censorship. In the United States, of course, personal data is being collected, monetized, and weaponized in ways that we are only just beginning to understand, and monopolies are squashing job creation and innovation. At this point it's harder and harder to argue that the benefits of platform technology vastly outweigh the costs.

Big Tech and big banks are also similar in the opacity and complexity of their operations. The algorithmic use of data is like the complex securitization done by the world's too-big-to-fail banks in the subprime era. Both are understood largely by industry experts who can use information asymmetry to hide risks and the nefarious things that companies profit from, like dubious political ads.

Yet that complexity can backfire. Just as many big-bank risk managers had no idea what was going into and coming out of the black box before 2008, Big Tech executives themselves can be thrown off balance by the ways in which their technology can be misused.[9] Consider, for example, the *New York Times* investigation in 2018 that revealed that Facebook had allowed a number of other Big Tech companies, including Apple, Amazon, and Microsoft, to tap sensitive user data even as it was promising to protect privacy.[10]

Facebook entered into the data-sharing deals—which are a win-win for the Big Tech firms in general to the extent that they increase traffic between the various platforms and bring more and more users onto them—between 2010 and 2017 to grow its social network as fast as possible. But neither Facebook nor the other companies involved could keep track of all the implications of the arrangements for user privacy. Apple claimed to not even know it was in such a deal with Facebook, a rather stunning admission given the way in which Apple has marketed itself as a protector of user privacy. At Facebook, "some engineers and executives . . . considered the privacy reviews an impediment to quick innovation and growth," read a telling line in the *Times* piece. And grow it has: Facebook took in more than $40 billion in revenue in 2017, more than double the $17.9 billion it reported for 2015.

Facebook's prioritization of growth over governance is egregious, but not unique. The tendency to look myopically at share price as the one and only indicator of value is something fostered by Wall Street, but by no means limited to it.[11] The obliviousness of the tech executives who cut these deals reminds me of bank executives who had no understanding of the risks built into their balance sheets until markets started to blow up during the 2008 financial crisis. Companies tend to prioritize what can be quantified, such as earn-

ings per share and the ratio of the stock price to earnings, and ignore (until it is too late) the harder-to-measure business risks.

Generation Greed

It's no accident that most of the wealth in our world is being held by
a smaller and smaller number of rich individuals and corporations,
who use financial wizardry like tax offshoring and buybacks to ensure that they keep it out of the hands of national governments. It's
what we've been taught to think of as normal, thanks to the ideological triumph of the Chicago School of economic thought, which
has, for the past five decades or so, preached—among other things—
that the only purpose of corporations should be to maximize profits.

The notion of "shareholder value" is shorthand for this idea.[12]
The maximization of shareholder value is part of the larger process
of "financialization," which I covered in my previous book, *Makers
and Takers*.[13] It's a process that has risen, in tandem with the Chicago School of thinking, since the 1980s, and has created a situation
in which markets have become not a conduit for supporting the real
economy, as Adam Smith would have said they should be, but rather,
the tail that wags the dog.

"Consumer welfare," rather than *citizen* welfare, is our primary
concern. We assume that rising share prices signify something good
for the economy as a whole, as opposed to merely increasing wealth
for those who own them. In this process, we've moved from being a
market economy to being what Harvard law professor Michael
Sandel would call a "market society," obsessed with profit maximization in every aspect of our lives. Our access to the basics—
healthcare, education, justice—is determined by wealth. Our
experiences of ourselves and those around us are thought of in
transactional terms, something that is reflected in the language of
the day (we "maximize" time and "monetize" relationships).

Now, with the rise of the surveillance capitalism practiced by Big Tech, *we ourselves* are maximized for profit. Remember that our personal data is, for Big Tech companies and others that harvest it, the main business input. As Larry Page himself once said, when asked "What is Google?": "If we did have a category, it would be personal information . . . the places you've seen. Communications . . . Sensors are really cheap. . . . Storage is cheap. Cameras are cheap. People will generate enormous amounts of data. . . . Everything you've ever heard or seen or experienced will become searchable. Your whole life will be searchable."[14]

Think about that, readers. You are the raw material used to make the product that sells you to advertisers. Yes, we really are living in the Matrix.[15]

Financial markets have facilitated the shift toward this invasive, short-term, selfish capitalism, which has run in tandem with both globalization and technological advancement, creating a loop in which we are constantly competing with greater numbers of people, in shorter amounts of time, for more and more consumer goods that may be cheaper thanks in part to the deflationary effects of both outsourcing and tech-based disruption, but that can't compensate for our stagnant incomes and stressed-out lives.

But you could argue that, in a deeper way, Silicon Valley—not the old Valley that was full of garage start-ups and true innovators, but the financially driven Valley of today—represents the apex of the shift toward financialization. Today the large tech companies are run by a generation of business leaders who came of age and started their firms at a time when government was viewed as the enemy, and profit maximization was universally seen as the best way to advance the economy, and indeed society. Regulation or limits on corporate behavior have been viewed as tyrannical or even authoritarian. "Self-regulation" has become the norm. "Consumers" have replaced citizens.[16]

All of it is reflected in the Valley's "move fast and break things" mentality, which the tech titans view as a fait accompli. As Eric

Schmidt and Jared Cohen wrote in an afterword to the paperback edition of their book, "Bemoaning the inevitable increase in the size and reach of the technology sector distracts us from the real question. . . . Many of the changes that we discuss are inevitable. They're coming."[17]

The Cost of Surveillance Capitalism

Perhaps. But the idea that this should preclude any discussion of the effects of the technology sector on the public at large is simply arrogant. There's a huge cost to this line of thinking. Consider that $1 trillion in wealth that has been parked offshore by America's largest, most IP rich firms. A trillion is no small sum: That's an eighteenth of America's annual gross domestic product, much of which was garnered from products and services made possible by core government-funded research and innovators. Yet U.S. citizens have not gotten their fair share of that investment because of tax offshoring. It's worth noting that while the U.S. corporate tax rate was recently lowered from 35 percent to 21 percent, most big companies have for years paid only around 20 percent of their income, thanks to various loopholes. The tech industry pays even less—around 11 to 15 percent—for this same reason: Data and IP can be offshored while a factory or grocery store cannot.

This points to yet another neoliberal myth—the idea that if we simply cut U.S. tax rates then these "American" companies will bring all their money home and invest it in job-creating goods and services in the USA. But America's biggest and richest companies have been at the forefront of globalization since the 1980s. Despite small decreases in overseas revenues for the past couple of years, nearly half of all sales from S&P 500 companies come from abroad. How, then, can such companies be perceived as being "totally committed" to the United States, or, indeed, to any particular country?[18] Their commitment, at least the way American capitalism is practiced today, is to customers and investors, and when both of them are increasingly global, then it's hard to argue for any sort of special consideration for American workers or communities in the boardroom.

Tech firms are *more* able than any other type of company to move business abroad, because most of their wealth isn't in "fixed assets" but in data, human capital, patents, and software, which aren't tied to physical locations (like factories or retail stores) but can move anywhere. And as we have already learned, while those things do represent wealth, they don't create broad-based demand growth in the economy like the investments of a previous era.

"If Apple acquires a license to a technology for a phone it manufactures in China, it does not create employment in the U.S., beyond the creator of the licensed technology if they are in the U.S.," says Daniel Alpert, a financier and a professor at Cornell University studying the effects of this shift in investment. "Apps, Netflix, and Amazon movies don't create jobs the way a new plant would." Or, as my *Financial Times* colleague Martin Wolf has put it, "[Apple] is now an investment fund attached to an innovation machine and so a black hole for aggregate demand. The idea that a lower corporate tax rate would raise investment in such businesses is ludicrous."[19] In short, cash-rich corporations—especially tech firms—have become the financial engineers of our day.[20]

The House Always Wins

There are the ways in which Big Tech is driving the mega-trends in global markets, as we've just explored. Then, there are the ways tech companies are playing in those markets that grant them an unfair advantage over consumers. For example, Google, Facebook, and increasingly Amazon now own the digital advertising market, and can set whatever terms they like for customers. The opacity of their algorithms coupled with their dominance of their respective markets makes it impossible for customers to have an even playing field. This can lead to exploitative pricing and/or behaviors that put our privacy at risk. Consider also the way in which Uber uses "surge pricing" to set rates based on customers' willingness to pay. Or the "shadow profiles" that Facebook compiles on users. Or the way in which Google and Mastercard teamed up to track whether online ads led to physical store sales, without letting Mastercard holders know they were being tracked.[21]

Or the way Amazon secured an unusual procurement deal with local governments. It was, as of 2018, allowed to purchase all the office and classroom supplies for 1,500 public agencies, including local governments and schools, around the country, without guaranteeing them fixed prices for the goods. The purchasing would be done through "dynamic pricing"—essentially another form of surge pricing, whereby the prices reflect whatever the market will withstand—with the final charges depending on bids put forward by suppliers on Amazon's platform. It was a stunning corporate ju-jitsu, given that the whole point of a bulk-purchasing contract is to guarantee the public sector competitive prices by bundling together demand. For all the hype about Amazon's discounts, a study conducted by the nonprofit Institute for Local Self-Reliance concluded that one California school district would have paid 10 to 12 percent more if it had bought from Amazon. And cities that wanted to keep on using existing suppliers that didn't do business on the retail giant's platform would be forced to move that business (and those suppliers) to Amazon because of the way that that deal was structured.[22]

It's hard to ignore the parallels in Amazon's behavior to the lending practices of some financial groups before the 2008 crash. They, too, used dynamic pricing, in the form of variable rate subprime mortgage loans, and they, too, exploited huge information asymmetries in their sale of mortgage-backed securities and complex debt deals to unwary investors, not only to individuals, but also to cities such as Detroit. Amazon, for its part, has vastly more market data than the suppliers and public sector purchasers it plans to link. As in any transaction, the party that knows the most can make the smartest deal. The bottom line is that both big-platform tech players and large financial institutions sit in the center of an hourglass of information and commerce, taking a cut of whatever passes through. They are the house, and the house always wins.

As with the banks, systemic regulation may well be the only way to prevent Big Tech companies from unfairly capitalizing on those advantages. Lina Khan, the antitrust lawyer we read about in chapter 9, explored this possibility in a *Columbia Law Review* paper[23] that argues that companies that both create marketplaces or platforms, and then also do commerce within them, have an unfair advantage. Her work calls in part upon that of a prescient academic, the Cornell University law professor Saule Omarova, who first came to my attention when I was researching *Makers and Takers*. Omarova was a key witness in hearings over the Goldman Sachs aluminum hoarding episode—as you might remember, the bank had found a diabolical way around rules saying that big financial institutions could not hoard raw materials like aluminum in order to drive up the price, but had to move the commodity in and out of warehouses to ensure that supply wasn't being interfered with. As a front-page *New York Times* piece exposed, Goldman was getting around those rules by simply using a forklift to move the aluminum back and forth between warehouses that were only a few feet away from each other.

Omarova's paper on the problem of financial institutions both owning and trading commodities, entitled "The Merchants of Wall Street: Banking, Commerce, and Commodities," sparked serious public interest in the topic. While the bank eventually offloaded its aluminum and came away from the episode without any legal action, Omarova said, "I'm sure that Goldman used the information they had about aluminum to influence the market." But, underscoring the opacity and complexity issue, she added, "Can I prove it? No. Can the CFTC [the regulator in charge of commodities trading] prove it? I doubt it. And if that's the case, should Goldman be doing any of this? Absolutely not."[24]

Omarova now believes that this dynamic is in play in the technology world, as big platforms both own the marketplace and trade within it. Her recent research raises questions about whether large tech platform firms pose the same kind of threat not only as the nineteenth-century railroads did (which also owned platforms and conducted business on them, as we've already learned) but also as the too-big-to-fail banks do.[25] She's particularly worried about the marriage of Big Tech companies and finance, and it's not hard to imagine why. "If Amazon can see your bank data and assets [what is to stop them from] selling you a loan at the maximum price they know you are able to pay?" Omarova asks.

She's not the only one worried. In June 2019, Christine Lagarde, the head of the International Monetary Fund, sounded the alarm about Big Tech, questioning whether the largest tech platform firms could destabilize the global financial system.[26] In December 2018, Agustín Carstens, the general manager of the Bank for International Settlements, spoke about the rise of Google, Alibaba, Facebook, Tencent, Baidu, eBay, and other companies in the global credit markets, calling them "one of the greatest challenges" to financial regulators today. "Will Big Tech's involvement in finance lead to a more diverse and competitive financial system, or to new forms of concentration, market power, and systemic importance?" he asked. "Is the expansion of Big Tech powered by efficiency gains? Or by the cost advantage of circumventing the current regulatory system?" It's a question that is ever more pressing as Facebook attempts to launch its own cryptocurrency.

The jury is still out on whether Big Tech will destabilize global finance. Meanwhile, Carstens and regulators in both the United States and Europe are looking carefully at whether the predictive algorithms and machine learning offered up by Apple, Amazon, Facebook, and others moving into the finance business are increasing or decreasing stability in the financial sector. One particular area of concern is how Big Tech firms use machines rather than human relationships to judge customers (thus circumventing many of the "know your customer" rules that govern traditional banking). As the mathematician Cathy O'Neil laid out in her book *Weapons of Math Destruction,* credit card companies and other financial institutions regularly use opaque algorithms that hoover up online data (our Web browsing patterns, for example, or our location data) and use them to create customer profiles that make it easier for people in upmarket zip codes that click on, say, a Jaguar during a car search rather than a Taurus, to get credit. That then creates a snowballing cycle of inequality; as she puts it, "A person using a computer on San Francisco's Balboa Terrace is a far better [credit] prospect than the one across the bay in East Oakland." That may, of course, be completely untrue. But the result is that what you do online may affect opportunities in your offline life in a big way.[27]

Then there are questions of whether Amazon or Facebook could leverage their existing positions in e-commerce or social media to unfair advantage in finance, using what they already know about our shopping and buying patterns to push us into buying the products they want us to in ways that are either (a) anticompetitive, or (b) predatory. There are also questions about whether they might cut and run at the first sign of market trouble, destabilizing the credit markets in the process.

"Big Tech lending does not involve human intervention of a long-term relationship with the client," said Carstens. "These loans are strictly transactional, typically short-term credit lines that can be automatically cut if a firm's condition deteriorates. This means that, in a downturn, there could be a large drop in credit to [small and middle-sized companies] and large social costs."[28] If you think that sounds a lot like the situation that we were in back in 2008, you'd be right.

Some people do not worry about any of this. They feel that such risks are a fair trade for the convenience of being able to link together all of our daily tasks, via a single password, on a single platform—be it Apple, Amazon, Google, or Facebook. But it is impossible to know what is "fair" when none of us can see inside the algorithmic black boxes of the largest technology companies. It is one thing for a company to know my vacation shopping patterns or what media I like; it is another for them to access my entire financial history, including my investments. Many people already lack confidence in making financial decisions and personal wealth management. Why else would so many of them still be paying above average fees for such services? Imagine how vulnerable some consumers might be if, for example, their bank notices that they have $9,000 sitting in a savings account and promptly serves them a pop-up ad urging them to move their money into a wonderful new higher yielding investment vehicle? Or imagine if your Facebook page had a checking account on top of it. What could go wrong?

It's a scenario we may face soon, given that the Trump administration has gutted the Consumer Financial Protection Bureau, which was established to protect ordinary people who have disputes with banks. Meanwhile, President Trump's Treasury is eagerly pressing for data sharing between tech firms and big banks to create "efficiency, scale and lower consumer prices." Again, we have to ask, are lower prices really worth more than all the hidden costs?

It's hard to know, thanks in part to the gutting of the Office of Financial Research, which is why there has been almost no study of the systemic risks and predatory pricing practices that could emerge when the world's largest technology firms and the biggest banks on Wall Street share consumer data. The lack of attention to these issues is no surprise, given the deregulatory stance that this administration has taken. But it's alarming nonetheless.

More alarming still is how some of the flow charts in the Treasury paper outlining how platform players and banks might work together to share consumers' financial information in order to offer "personalized" products and services remind me of the complex illustrations of credit default swaps that we saw in the wake of the 2008 crisis. Both are Rube Goldberg–style studies in risk. Complexity of that sort always makes me nervous, as it leaves so much room for the party with more information to obfuscate. Call me a Luddite, but I've always agreed with former Fed chair Paul Volcker that the ATM has been the most useful "innovation" in finance in the past few decades.

Too Big to Regulate?

No matter what the Silicon Valley giants might argue, size, ultimately, is a problem, just as it was for the too-big-to-fail banks. This is not because bigger is inherently bad but because the complexity of these organizations makes them so difficult to police. Like the big banks, Big Tech uses its lobbying muscle to try to avoid regulation. And like the banks, it tries to sell us on the idea that it deserves to play by different rules.

It does not. If anything, I'd argue that when firms become this big and this complicated, they need many more rules than most other companies. At the very least, Facebook, Google, Amazon, and the other systemically important platforms should be forced to disclose political advertising in the same way that television, print, and radio firms do. When in the financial markets, they should be forced to stay in their own sandbox the same way that their competitors, the big banks, are. When trafficking in other sorts of personal information, like, for example, healthcare information (which can be collected and sold via any number of health and wellness apps that we all use), they should be bound by the same privacy rules that the healthcare industry follows. Big Tech isn't special. But it is systemically important. It sits in the middle of communications, media, and advertising markets, in the same way that big banks sit in the middle of financial markets. The Fed or the FTC can step in when structurally important financial institutions are both making markets and participating in them; the FTC should do the same with Big Tech. In fact, I'd argue that we need a specific regulatory body for the technology industry, a topic I'll come back to in the last chapter.

Treating the industry like any other would undoubtedly require a significant shift in the Big Tech business model, one with potential profit and share price implications.[29] The extraordinary valuations of the Big Tech firms are due in part to the market's expectations that they will remain lightly regulated, lightly taxed monopoly powers.[30] But that's not guaranteed to be the case in the future. Antitrust and monopoly issues are fast gaining attention in Washington, where the titans of Big Tech may soon have a reckoning.

CHAPTER 11

In the Swamp

B ack in 2012, University of Maryland law professor Frank Pasquale, a scholar in the area of information law, accepted an invitation to speak at what he thought was a straightforward academic conference at George Mason University on the topic of competition, search, and social media. At the time, Google was under investigation for anticompetitive behavior in the search market in the United States (where the FTC was investigating the Yelp case) and in Europe (where the British comparative shopping start-up Foundem was challenging the search giant on what it felt was the use of discriminatory price-comparison algorithms). The legal challenges raised by the burgeoning digital economy were becoming more pressing by the day, and Pasquale was eager to exchange views on the changing landscape with other academics.

Pasquale, a tall, lanky, and affable information law scholar who bears a passing resemblance to a young Jimmy Stewart in both looks and demeanor, had been researching Google and the other tech giants for some time. His 2015 book, *The Black Box Society,* had yet to be published. But he had already become a go-to source for scholars, businesspeople, and policy makers who wanted to better understand how the platform tech giants were collecting our personal data, then connecting the dots of our behavior and selling those insights to the highest bidder.

His presentation, "Search, Antitrust, and Competition Policy," focused on Google, specifically how the targeting of personal data and analysis of our every movement online gave the firm enormous power to both help and hurt consumers.

"I shopped yesterday at a big and tall menswear shop, and today, I'm being given ads for airline seats with more legroom," said Pasquale, who outlined in detail the way in which Google's surveillance capitalism worked. He compared the policy makers' obliviousness of the implications of these practices to the way in which regulators were willfully blind to the risks posed by the financial sector in 2008. As he later put it in his book, "Lobbyists for the black box industries mock the capacity of government to comprehend the business practices of a Google or a Goldman." Yet, as he notes, there are plenty of examples of complex industries—consumer products, pharmaceuticals, and healthcare, for starters—that are successfully regulated, to the benefit of all.[1] These industries also spend a lot of money on lobbying, but somehow regulators still have been able to tame them, often after a crisis prompted public outrage.

But unbeknownst to Pasquale, he was just the opening act in a panel in which the deck was stacked against him. Shortly after he spoke, Scott Sher, an attorney from Google's own law firm, Wilson, Sonsini, Goodrich, and Rosati, took the stage. But rather than an academic presentation, he delivered what amounted to a brief of Google talking points ("Competition is a click away," and the like). He also laid out harsh refutations of Pasquale's critique of Google's monopoly power, claiming that a 65 percent market share (which is what the firm had at the time) in search was nothing to be concerned about.

Pasquale felt completely blindsided. Ordinarily at scholarly conferences, academics present their own research, and debate back and forth. Sher essentially gave a PR presentation for Google. "There was no time at the end for me to respond, though I had listed about fifteen things I planned to dispute in his presentation," says Pasquale. "I wish I had simply demanded the floor, but that risked making me seem like a wild-eyed radical."

Only later did Pasquale find out that he had been set up: He was a token Google critic, capable of giving some semblance of balance to a conference that was largely planned, bought, and paid for by Google itself.[2] As *The Washington Post* later reported, it was part of a campaign by Google executives to shift the antitrust conversation in the nation's capital, at a time when the FTC was, as we've already discovered, seriously considering whether it should break the company up. As an email trail shows, the executives had "suggested" mostly speakers who they knew to be sympathetic to Google, and worked with the GMU Law and Economics Center to structure panels in such a way that the company would appear in a positive light. They also made sure that key FTC personnel like Beth Wilkinson, a Washington litigator who had been appointed to head up the investigation into Google, received invitations.

Google's "Silicon Tower"

I was sympathetic to Pasquale's story for many reasons, one of which is that I myself have been in similar situations. In 2017, I appeared on the BBC's *Intelligence Squared* program to debate the subject of whether Big Tech firms should be broken up as anticompetitive monopolies. I was speaking for the motion, and University of Leeds professor Pinar Akman vigorously argued against such measures. She had her own legitimate arguments to make. But it's worth noting that a chunk of her research is paid for by Google.[3] In studying the tech sector, I have also, in recent years, turned to the work of various academics, like, for example, Stanford Law professor Mark Lemley, a particularly prolific publisher of work in which Google might have an interest. I eventually found in the fine print in the footnotes of some of his research that he was also a Google "consultant."[4]

I don't mean to pick on these academics particularly. As it turns out, all this is just the tip of the iceberg. Over the past decade or so, Big Tech in general but Google in particular has bought and paid for mountains of academic research conducted on areas of interest to the firm. A July 2017 *Wall Street Journal* investigation into the topic found that Google had financed hundreds of research papers to defend against regulatory challenges to its market dominance, shelling out anywhere between $5,000 to $400,000 to a host of academics, consultants, and former or future government officials.[5]

Some of the money flowed to Daniel Sokol, a University of Florida professor who in 2016 published a paper arguing that Google's use of data was perfectly legal. At the time, he noted that no companies had funded the research. What he didn't note was that he himself was a part-time attorney at Wilson Sonsini, and that his paper's coauthor was a partner at the law firm, too. Nor did Sokol note that in March 2013, he had helped Paul Shaw, a Google public policy official (Google has dozens of them, along with hundreds of PR staff and legions more lawyers, all of whom work together to push issues of interest), persuade law professors to write papers sympathetic to Google's points of view for an online symposium on patents. The professors weren't paid. But Sokol did submit a $5,000 bill to Google for his work.

In 2016, two groups—the Google Transparency Project and the Campaign for Accountability—known to be go-to sources for anyone looking for information on Google's funding of academics and government officials, put out a paper entitled "Google's Silicon Tower," looking at the presence of Google-funded speakers at key 2016 policy conferences arranged by the Federal Trade Commission, George Mason University, and Princeton. Among the findings: More than half of the speakers at the FTC's 2016 PrivacyCon (twenty-two out of forty-one) were funded by Google, and more than half of the research papers presented there had an author with financial ties to Google—only one of whom had disclosed that funding.

Perhaps more shocking was that Lorrie Cranor, the FTC's chief technologist at the time, had received $850,000 dollars from the search giant, including some $350,000 in personal research awards and $400,000 in shared grants. And at George Mason's event on global antitrust investigations of Google, four out of five speakers had received funding from none other than Google, while five out of seven of the panelists at Princeton University's broadband privacy workshop received support from the company.[6]

It's important to note that the "Silicon Tower" paper was paid for in part by corporations that have their own various legal or competitive battles with Google. (The Google Transparency Project, for example, is backed by Microsoft and Yelp, among others, and the Campaign for Accountability, which doesn't disclose corporate funding, has backed various Oracle positions over the years.) But that fact doesn't make the findings untrue, just as the research funded by Google itself isn't necessarily without any merit (although it's hard to imagine that the researchers doing it aren't somewhat biased in favor of the company). Both sides have their arguments to make. But what all this tells us is how the debate around key issues of economic policy in the most important industries in our country has been almost entirely hijacked by large corporations with deep pockets.

The fact is that the public debate around monopoly, privacy, cybersecurity, and so forth (to the extent that we are even having a public debate) is largely orchestrated by the very companies upon which the debate centers. These aren't payoffs per se; "I know from interacting with many of the academics in this Google firmament that it's mostly not about a specific quid pro quo. It's much more subtle than that," says one senior aide for an influential Democratic senator working on legislation around these issues. "It's about social and intellectual capture, which is actually much more effective, both short- and long-term. Google supports researchers working in areas that are complementary to Google business interests, and/or adverse to its competitors' business interests; things like relaxed copyright laws, patent reform, net neutrality, laissez-faire economics, privacy, robots, AI, media ownership, government surveillance (which is often villainized in order to draw attention away from extensive corporate tracking), et cetera. They do this via direct grants to the researchers, funding of their centers and labs, conferences, contributions to civil society groups, and flying them out to Google events."[7]

In this way, the company not only builds goodwill, but successfully "grooms academic standard-bearers, prominent academics who will drive younger peers in a direction that is more favorable to the company," says the aide. It seems that truly independent academics like Frank Pasquale are, these days, almost impossible to find.

Follow the Money

All the money is, of course, about controlling the policy debate in Washington. Whenever there are moves by politicians to rein in Big Tech, the companies can trot out their paid-for experts. Witness the congressional hearings on monopolies in 2019 following Democrat Elizabeth Warren's calls to break up Big Tech firms.[8] The experts giving testimony included Joshua Wright, a former Trump adviser and professor at George Mason University, who had written academic research funded indirectly by Google, and criticized antitrust scrutiny of Google shortly before joining the Federal Trade Commission, after which the FTC dropped their antitrust suit.

Senator Josh Hawley, a Republican who had brought a case against Google during his tenure as Missouri's attorney general, asked how anyone could possibly justify not acting to regulate Big Tech when "every day brings some creepy new revelation" about everything from the misuse of consumer data to cronyism. A *Wired* article quoted Wright as saying simply that his views had "attracted like-minded supporters" as the antitrust debate had grown.[9]

No wonder. Over the past few years, the technology industry has quietly become the dominant political lobbying power in America, in terms of the sheer amount of cash and soft power it exerts in an effort to avoid regulatory disruption of its business model. According to the Center for Responsive Politics, the Internet and electronics industry together spent a record $216.4 million on federal lobbying in 2017 and $224.6 million in 2018—more than any other industry except Big Pharma.[10] Google was the highest spending corporate lobbyist in both 2017 and 2018.[11] And Amazon takes the cake in terms of the sheer number of issues it lobbies, disclosing that it lobbied forty federal agencies on twenty-one different general issue areas in 2018. Some of the topics are the usual ones that Big Tech firms lobby around: net neutrality, telecoms, and data standards. But others were much more Amazon specific: The company weighed in on issues related to self-driving cars and drones, which it plans to use to carry its massive yearly freight of goods; it also lobbied to allow groceries to participate in certain government programs (reflecting its new interest in that area via its Whole Foods acquisition), and pushed for laws that would be more favorable to its pharmaceutical business (the company bought an online drug store, PillPack, in 2018).[12] Even Netflix, which is in some ways the Teflon FAANG, has major lobbying efforts going in both the United States and Europe around issues like copyright, privacy rules, and various digital regulations.[13]

That's only the money we can see. Silicon Valley also funds any number of unrelated nongovernmental organizations and interest groups that then help argue their case, either explicitly or simply by abandoning agendas that could be harmful to tech. Google alone has thrown money at more than 140 such third-party entities, including various nonprofits, academic institutions, and media fellowships. The media capture I find particularly upsetting, for obvious reasons. Facebook, in the wake of the 2016 election-meddling scandal, decided to start "paying indulgences to make up for some of its sins against journalism," as a 2019 *Wired* cover story on the company put it,[14] by throwing hundreds of millions of dollars toward supporting the local news industry, which had itself been decimated by the platform giants, which have taken nearly all new digital ad dollars. Google also donates large sums of money to news organizations for the development of certain types of content[15] (many brand-name publications have been the recipients of such "largesse" from the platform giants). You could argue that this shows that Big Tech is trying to repair the damage that it has wrought to real news, and to liberal democracy in general. I'd argue the opposite—journalism wouldn't be in this sort of shape if content creators had received a fair share of the revenue for what they create from the very beginning.

Cognitive capture is a subtle art. But, if you follow the money long enough, it will lead you straight through the revolving door between Washington and Silicon Valley, whereby Google, Facebook, and others regularly hire influential former government officials who then move in and out of business and policy circles, pushing tech-friendly notions like the idea that privacy is somehow a civil liberty infringement, or that cheaper prices should be the key metric for consumer goods, or that addictive technologies are actually good for kids.

Google isn't the only firm with a revolving door, though it has had a substantial one over the past few years. Plenty of other tech companies have aspired to this sort of influence. Uber, for example, hired David Plouffe, the man who helped Barack Obama reach the White House, to run its communications and political work in 2014. Following Plouffe's hiring, Uber picked up better lobbyists, academic research to back up its positions, and an endorsement from Mothers Against Drunk Driving.[16]

Facebook, too, has hired dozens of former politicos to join its lobbying operations, including Jeff Sessions's former legislative director Sandy Luff, Nancy Pelosi's former chief of staff Catlin O'Neill, and longtime John Boehner aide Gary Maurer. The Facebook and Google revolving doors are especially alarming given that politicians are customers of the two companies, in the sense that the social media and search sites have been such a huge part of campaign efforts on both the right and the left (more so with every passing election).[17]

As academics Daniel Kreiss and Shannon McGregor documented in their paper "Technology Firms Shape Political Communication: The Work of Microsoft, Facebook, Twitter and Google with Campaigns During the 2016 U.S. Presidential Cycle," this is work that's been going on for some time now. Not only do such companies sell their services to political campaigns (making loads of money in the process), but they also actively shape campaign communications, acting as "quasi digital consultants . . . shaping digital strategy, content, and execution."[18] Far from being neutral platforms or even just traditional media players, Big Tech has moved into the political consultant space, becoming "active agents in political processes."

"The Biggest Kingmaker on Earth"

While far from the only tech company engaged in a massive campaign for Washington mindshare, it was Google, whose executives visited the White House more than any other corporation's during the Obama years, that had by that time become "the biggest kingmaker on this earth."[19] How they have wielded that influence underscores the way in which money in politics has completely distorted our economy, undermining both competitiveness and public trust in institutions.

Consider the issues of data privacy and antitrust, for example. One of the major turning points for Google on those issues was the acquisition in 2007 of the ad network DoubleClick, which was the leading firm that helped advertisers and ad agencies decide which websites would be best for hosting their ads. As Steven Levy writes in *In the Plex,* "the DoubleClick deal radically broadened the scope of the information Google collected about everyone's browsing activity on the Internet."[20] Competitors and regulators alike questioned the deal, which eventually went through, in large part because Chicago School thinking didn't really leave any room for a good antitrust argument against it (despite the fact that it would allow Google to essentially control the vast majority of advertising online). But Google questioned it, too, at least in terms of the ramifications for its own bottom line. Larry Page and Sergey Brin were at first reluctant to combine the data and information that could be harvested via cookies on its own platform with what could now be garnered via DoubleClick (which was, of course, now a part of Google itself). But eventually, under pressure to grow, the company relented. Thanks to the merger, "Google became the only company," writes Levy, "with the ability to pull together user data on both the fat head and the long tail of the Internet. The question was, would Google aggregate that data to track the complete activity of internet users? The answer was yes."[21]

While Google had long promised users that it would ask their permission if it ever used their data for anything other than the purposes for which they'd given it (that is, for whatever individual search, or email, social media, or map functions they'd signed up for), it had begun combining and selling all the data it had on users to the highest bidder. So much for worrying about "advertising and mixed motives," as Page and Brin had put it in that not-so-long-ago 1998 paper in which they fretted about the way that a search engine funded by targeted advertising might possibly do harm. The new way of handling user data wasn't clearly in violation of U.S. law, but it wasn't exactly transparent, either, and as it turned out, parts of the Google effort did violate the U.S.-EU Safe Harbor Framework, which governed how data could be transferred between the two regions. This was a fact that became evident after Google launched its short-lived social networking service, Buzz, in 2011, and came under fire for the way in which it was combining and releasing user information.[22]

In March 2011, the FTC issued a complaint and consent order requiring Google to change the way it handled data; it directed the company to get explicit consent from users before disclosing such data to advertisers (or any third party).[23] Yet less than a year later, in a blog post, the company boasted that it would "treat you as a single user across all [its] products, which will mean a simpler, more intuitive Google experience."[24] A nonprofit watchdog group, the Electronic Privacy Information Center (EPIC), filed a complaint calling on the FTC to disallow this consolidation of personal information and shortcut around privacy.[25]

A raft of user lawsuits predictably followed. But the courts were unable to compel the FTC to enforce its own action, and over the next few years, the company touted the ways in which it was combining and monetizing surveillance across a broader number of apps, platforms, and devices, even claiming it would soon be able to track conversations involving store visits and phone calls.[26] After more complaints, in 2012, the company paid out the largest FTC civil penalty ever to settle charges that it had exploited a loophole to bypass the privacy settings of Apple's browser Safari and had tracked users of the browser, while giving the impression that they weren't being monitored.[27]

Here, it's important to point out that for years in advance of these actions, Google had been quietly lobbying behind the scenes for support on its positions. That involved, as per usual, throwing a lot of money into universities, think tanks, and nongovernmental groups that might eventually be called on to back those positions publicly. In a 2019 *Wired* piece looking at the way in which Google influences the conversation in Washington, Georgetown law professor Marc Rotenberg, the president of EPIC, noted that Google did exactly that after EPIC filed complaints about the company's acquisition of DoubleClick in 2007 and Nest in 2014. "Money buys silence," he said. "Google doesn't need the experts to agree with them. They only need them to look the other way."[28]

As for the regulators themselves, it's hard to argue that they were fully doing their jobs. Large fines as the cost of doing business had become status quo for Big Tech, just as they are for Big Finance. Silicon Valley was learning that it could simply shell out cash from its growing coffers, and regulatory problems would go away. It was no secret that the people who were supposed to look out for citizens' interests were essentially unable or unwilling to effect any real change in corporate behavior. In fact, in many cases, it seemed that the interests of the companies and the policy makers were quite closely aligned. As one Senate aide ruefully put it to me during the reporting of this book, "Google was Obama's Halliburton."

As Google's influence in the Obama White House grew, the company developed a growing interest in the controversial subject of a big report on the future of the Internet written by a commission chaired by Google VP Marissa Mayer that was released by the White House in 2009.[29] The paper advocated for "net neutrality," a term coined in the early 2000s by the academic Tim Wu, who has since become a vocal Big Tech critic.

Net neutrality was at the heart of a growing divide between the digital giants and major telecom companies over "open-access" policies. The Big Tech platform players supported net neutrality legislation that prohibited network providers such as AT&T, Verizon, and the like from charging more to prioritize certain kinds of content. Google, for example, wanted to make sure that Comcast couldn't run more Hulu content than YouTube content if Hulu was willing to pay more. It was essentially a corporate fight for turf. But Big Tech lobbyists and policy makers cleverly branded it as a fight for the little guy.

Net neutrality came to be understood as shorthand for the idea that everyone—rich and poor, the start-up and the multinational conglomerate—should be able to use the Internet on a level playing field. Liberals in the United States supported the idea for obvious reasons of social equity. But some conservatives, as well as some members of the business community, argued that it would prevent the Internet providers from properly monetizing their investments in broadband (which they, not Google and its ilk, were making). It was a fair point. After all, the telecom companies building the twenty-first-century digital highway have single-digit profit margins, while the likes of Google and Facebook—which simply have to wait for someone to upload a cat video and then sell hypertargeted advertising against it—have profit margins in the high double digits.

What was lost, of course, was the middle ground. Certainly, it makes sense that individual Internet users should pay the same data rates, regardless of the content they were consuming. There should be no rich-poor divide for citizens online. There's also no reason that the rights of individuals should be a casualty of turf wars among big corporations. Simply set different rules for individuals versus large companies. And yet, Big Tech–funded groups like the Electronic Frontier Foundation (which certainly has many good civil liberty intentions behind what it does, but also receives millions in funding from the net giants that have been the largest corporate beneficiaries of net neutrality) have argued vociferously that cable providers shouldn't be able to charge, say, Google or Netflix a bit more for all the massive downloading of video done via those sites, and that more power for the Internet service providers would squash innovation on the Internet and unfairly penalize small businesses. A number of critics would argue that Big Tech companies *themselves* are a bigger risk to innovation than the telecom companies, in large part because of the network effects that make them natural monopolies and enable them to become ever more dominant and able to squash competition in myriad ways.

In 2015, the FCC issued the Open Internet Order, which allowed the tech companies to do just that. It forced the cable companies to offload the costs of "freedom" to users, one result of which has been a slowdown in broadband deployment, particularly in rural areas.[30] A few years back, Google itself tried to increase penetration in underserved areas in various parts of the United States, but slowed efforts after it found that laying fiber was much harder and less profitable than monetizing personal data via targeted advertising.[31]

In 2017, the FCC under President Trump rolled back net neutrality rules. House Democrats, in 2019, passed a bill reinstating the rules and are keen to make this an important election issue. But there has yet to be a clear and honest debate about whose interests are really being looked after—consumers' or big companies'—on both sides.[32] Telecom critics have argued that since the rollback in 2017, big cable companies have not actually invested more in the capital spending that they promised to make to improve networks.[33] That's a very fair point, and one that calls into question some of the original telecom arguments around net neutrality (though there are plenty of other reasons to slow capital spending right now, like the risk of a looming recession and controversy over 5G rules and regulations). But it's also true that broadband speeds have increased since 2017, and there doesn't yet seem to be any sort of two-speed Internet, or rich-poor divide in terms of how quickly you can download those cat videos.[34] The point here is that this fight isn't really so much about public interest as it is relative corporate welfare.

Net neutrality is one area in which the public debate over an issue that is central to our economy and civic society has been captured. Patents, as I covered extensively in chapter 5, is another. Over the past few years, I've heard disparate complaints from a variety of quarters—from start-up biotech firms, semiconductor and electronics firms, clean-tech companies, data analysis groups, universities, and innovators working on the Internet of things, as well as some of the venture capitalists that invest in these areas—that the patent system and the debate about how to structure it has been hijacked by the interests of the largest tech firms in the country. Indeed, the only ones who seem not to be complaining about the current system are Google, Apple, Intel, Cisco, and other Silicon Valley giants.

While they all have their own patents to protect, their business models, which involve products that include hundreds or even thousands of bits of intellectual property, tend to do better when there are fewer patents to deal with. But small and mid-sized software and hardware suppliers, as well as life sciences companies, have different business models—ones that live or die on the ability to protect a handful of patents, and thus monetize their years of investment. They believe that the patent pendulum has swung too far. (A bipartisan group of members of Congress as well as the current head of the U.S. Patent and Trade Office feel that it probably has, though they've had trouble getting too much traction for those ideas.)

Big Tech would, of course, disagree. But the point here isn't that this is a one-sided debate, but rather how easy it is for the largest corporate interests to change the game to suit themselves in ways that damage the economic ecosystem as a whole. There are any number of other areas we could look at that would illuminate the way in which Big Tech has bought out the public debate, from regulation of self-driving cars to public surveillance to antitrust to copyrights. But to be fair, oftentimes, the advantages leveraged by Big Tech come down to legal inertia as much as active lobbying for loopholes. In the absence of smart regulation, the industry continues to benefit from rules that were set down decades ago, when modern tech companies were just getting off the ground.

In fact, many of the key laws that govern digital commerce (which, increasingly, is most commerce) were crafted in the 1980s and '90s, when the Internet was an entirely different place. Consider Section 230 of the Communications Decency Act, which gave technology companies exemption from liability for what people do and say on their platforms. In 1996, when this law was crafted, no one could have predicted that it would come to serve as a legal loophole for companies like Backpage.com that deliberately created—and profited handsomely from—a platform for online sex trafficking. On August 1, 2017, a bipartisan group of senators, led by Democrat Claire McCaskill of Missouri and Republican Rob Portman of

Ohio, introduced legislation that would create a carve-out in section 230 for tech firms that knowingly facilitate sex trafficking, meaning they could be held responsible for that. The bill was called SESTA, the Stop Enabling Sex Traffickers Act. It was the rare piece of legislation that everyone, it seems, could get behind—except the largest tech companies and their industry lobbying groups, who were concerned that it would open a Pandora's box of legal issues for them.[35]

So, they decided to fight it. These groups had the rough copy of the bill for months before its introduction, yet refused to offer edits during its crafting. "We did our due diligence, met with the tech community on a bipartisan basis for months, and yet they offered no constructive feedback," said Kevin Smith, a spokesperson in Portman's office. The tech companies made it clear that amendment to 230 was a no-go, and even suggested a number of (weak) alternatives, like tougher criminal laws. As Noah Theran, a spokesperson for the Internet Association, a trade group that represents Google, Facebook, and other companies, told me at the time, "The entire Internet industry wants to end human trafficking. But there are ways to do this without amending a law foundational to legitimate Internet services."

As one might imagine, this attitude didn't make for very good press for the industry, particularly after *New York Times* columnist Nicholas Kristof's piece decrying Google's use of lobbying power to push back against legislation that would crack down on sex trafficking.[36] Finally, Facebook, which was looking to build some goodwill following months of pressure over its role in the 2016 election manipulation by Russia, caved and decided to endorse the bill. The Internet Association eventually caved, too. Google remained quiet, perhaps because of a damning Consumer Watchdog report that outlined the way in which it had, over a long period of time, been covertly fighting against congressional efforts to close the CDA 230 loophole, despite the effects on sex trafficking.[37] Two other key lobbying groups, the Consumer Technology Association and NetChoice, also continued to oppose the legislation.[38] The law, which was a no-brainer, eventually passed.

But the industry continues to try to find backdoor ways to reverse or circumvent it, such as through the renegotiation of trade deals.[39] The Electronic Frontier Foundation challenged SESTA's constitutionality, despite proof that it helped reduce sex trafficking and online crime.[40]

To be completely fair here, the EFF's worries about the slippery slope of censorship are legitimate. They weren't wrong to be concerned about the notion of platform firms being asked to police the Web given that, as their policy director, David Greene, told me in 2018, "they are so terrible at it."[41] You can certainly see it from both sides. But leaving things as they are isn't really a solution, either. These problems aren't going away—and indeed, both the current lack of liability that the platform firms enjoy and the fact that they aren't particularly good at moderating problematic content underscore the fact that a new regulatory framework for the digital economy is desperately needed. And yet, because politicians—not to mention nonprofit think tank experts, academics, and even some journalists—have been so cognitively captured by the usual Big Tech narrative, that hasn't yet happened.

Here, the fiction they have invented about Europe is instructive. Many American businesspeople and some politicians rail against "statist" old Europe, painting it as a sclerotic place that illustrates why regulation is the enemy of growth. I've heard from a number of European policy makers how tired they are of CEOs from Silicon Valley "coming over here and telling us we aren't innovators, we didn't create the biggest tech companies, and our concerns [about privacy and monopoly] are just sour grapes."[42]

But a fascinating study by academics Germán Gutiérrez and Thomas Philippon showed that EU markets are, in fact, more competitive by many measures than those in the United States, and points to a huge rise in U.S. political lobbying as the key reason that the European Union has significantly lower levels of corporate concentration, excess profits, and regulatory barriers to entry since the 1990s.[43] The study found that this jump in dollars spent on U.S. political lobbying is the key reason that levels of concentration between the two regions have diverged since the 1990s.

"European institutions are more independent than their American counterparts, and they enforce pro-competition policies more strongly than any individual country ever did," Gutiérrez and Philippon wrote. If you look at competitiveness in terms of GDP numbers and the size of the biggest companies, Europe lags behind the United States. But as most economists today would agree, GDP alone is an inadequate metric for well-being, and more corporate concentration is a sign of poor economic health rather than vibrancy. This serves as a sharp counterpoint to the argument often used in Silicon Valley that Europeans do not have an Internet giant because they simply are not innovative. It's true that Europe doesn't have the equivalent of the FAANGs. But by some measures, it's all the better for it.

None of this is to say that Europe doesn't have its own problems with regulatory capture and unhealthy ties between government and the private sector. The dysfunctional German banking system, for example, is a perfect illustration of how cozy ties can create conflicts of interest that ultimately backfire—recall the highly leveraged German state-owned banks that blew up in 2008, and the ongoing Deutsche Bank bailout saga today.

But in the United States, the sheer amount of political power exerted by just a handful of big companies is breathtaking. Consider, for example, Google's 2019 announcement of plans to roll out faster Web content for Cuba,[44] which would seem to counter President Trump's promises to get tough on the regime of Nicolás Maduro in Venezuela. The Cubans, you see, are understood to provide Caracas with intelligence services. How was it that Google in particular got first dibs to wire Cuba? Because former chairman Eric Schmidt and three other executives at Google—which had tremendous pull with the Obama administration as we've already learned—flew into the country in the midst of an embargo in 2014. Six months later, U.S. policy toward Cuba changed. Now Google may send YouTube videos to the island under Trump.

It's not a huge deal commercially for Google—and some would argue it's better than Latin oligarchs doing it. But it encourages the conclusion that the United States is more like Latin America than many of us would like to think: a place where money and personal connections matter more than anything else. No wonder millennials see markets as being politically constructed in ways that their parents did not. While they are too young to remember the failures of communism, they see the hypocrisies of capitalism all around them.[45]

It's useful to remember the last time a big industry captured the regulatory system in ways that undermined both economic security and public trust: the years before and after the 2008 financial crisis. One reason that the Dodd-Frank financial reforms and regulations took years to write and arguably didn't end up really addressing the core issues of making our markets safer is that there were too many vested interests in the room. Banks themselves were involved in 90 percent of the consultation meetings about the most contentious bits of regulation, and between the lobbyists, the seven financial regulatory bodies, and the various political factions, the resulting legislation had more holes than Swiss cheese. This is one reason it's been relatively easily for the Trump administration to roll back some of Dodd-Frank—there were plenty of people on both the right and the left who didn't like it to begin with because it had been made so complicated, and in some cases so ineffectual, by industry lobbyists themselves.

We are seeing the same pattern today with Silicon Valley. One can only hope it doesn't add kerosene to the fire of populism that was lit by the 2016 elections.

CHAPTER 12

2016: The Year It All Changed

If Facebook were a country, it would be the biggest one on the planet. Over 2 billion people, more than a third of people living today, log in every month. It knows more about us than pretty much anyone else, barring our closest friends and family, who are, of course, right there, every day, on the platform with us. When you think about that, it was only a matter of time before nefarious actors found a way to exploit that data to corrupt the democratic process.

Politicians have for years been using detailed data metrics and market feedback to try to influence election results. But it wasn't until 2016 that the general public really understood the vast possibility held by such techniques when amplified by the kind of surveillance capitalism practiced by the largest platform technology firms. In the run-up to the presidential election, political groups spent a total of $1.4 billion in online advertising and marketing, up four times from the last election cycle.[1] Facebook and Google were, of course, both huge beneficiaries. But they were also involved in yet another way: They "embedded" employees (a term that PR representatives have since tried to eliminate from media reports, but which perfectly captures the depth and breadth of their involvement) with the Trump campaign, essentially providing free staff to help politicos figure out how best to use the platform to get their message across to potential voters.

While the ultimate results of that were unprecedented, the practice itself was not; media organizations have often worked closely with political campaigns. As reams of research and reporting over the past few years have shown, the Big Tech companies—not just Facebook and Google but others like Twitter and Microsoft and Apple as well—have taken political communications to an entirely new level. Since 2012, these companies have been intimately involved in campaign strategy on both sides of the aisle. Google and Facebook staff both collaborated with the 2012 Obama and Romney campaigns to orchestrate digital advertising purchases. In 2014, Twitter released a 136-page manual on how to use the platform to influence voters in elections. Also in 2014, a strategy memo sent to Hillary Clinton's campaign chairman and other senior leadership outlined how the Clinton campaign might work with tech firms. "Working relationships with Google, Facebook, Apple and other technology companies were important to us in 2012 and should be even more important to you in 2016, given their still-ascendant positions in the culture. These partnerships can bring a range of benefits to a campaign, from access to talent and prospective donors to early knowledge of beta products and invitations to participate in pilot programs."[2]

But 2016 was the year that Big Tech really blew up the old style of political campaigning. At the 2016 Democratic National Convention in Philadelphia as well as at the 2016 Conservative Political Action Conference (CPAC) in Maryland, Big Tech firms had a huge presence. In Philly, Google took over an entire industrial space where politicians and staff could mingle with tech workers. The company also worked with Republicans like Rand Paul, whose digital director Vincent Harris flew out to Google headquarters for "ideation" sessions on campaign content and advertising. Facebook had prime real estate, too, at both conventions, and also worked "to teach conservative-leaning candidates [at CPAC] how to use its platform to reach new voters."[3]

The relationship was rich for both sides. As academics Daniel Kreiss and Shannon McGregor laid out, "Firms are motivated to work in the political space for marketing, advertising revenue, and relationship building in the service of lobbying efforts. . . . Facebook, Twitter and Google go beyond promoting their services and facilitating digital advertising buys, actively shaping campaign communications." The academics concluded that "these firms serve as quasi-digital consultants," shaping "strategy, content and execution."

And from the point of view of the campaigns, why not? Big Tech firms could serve as "free labor," which is how Nu Wexler, associate communications director at Twitter in 2017, put it in an internal communication. Wexler, who was interviewed by Kreiss and McGregor, noted that the Trump campaign in particular compensated for its small staff by using the expertise of tech firms as advisers on strategy and communications. "The Trump model was that they . . . rented some cheap office space out by the airport, a strip mall, and they said it's going to be Trump Digital. They had the companies, the advertising companies, social media companies, come down there [to San Antonio, the campaign headquarters] and work out of that strip mall. And we did it, Facebook did it, Google did it." Much of the help, according to Wexler, was around building "ads that get results."[4]

Results are one thing. But the dirty politics of the Trump campaign, coupled with the vast trove of data that could be harvested from Facebook and a variety of other websites and apps, led to something much darker.[5] According to a report published in *Bloomberg Businessweek* that came out a full two weeks before the election, Trump knew he needed a miracle to win, and his campaign team, led by the infamous Steve Bannon; Trump's son-in-law, Jared Kushner; and other social media–savvy staffers, found that "miracle" in Facebook, Twitter, and YouTube.[6]

232 DON'T BE EVIL

Kushner, who had friends in the tech world, reached out to "some Silicon Valley people who are kind of covert Trump fans and experts in digital marketing." These techies showed the campaign how it could target particular groups of swing voters, either to bring them out or to suppress them, by sending out targeted online messages in advance of the elections. The *Businessweek* story quoted a senior official in the campaign who admitted "we have three major voter suppression operations under way. They are aimed at three groups Clinton needs to win overwhelmingly: idealistic white liberals, young women, and African Americans."

The campaign succeeded in throwing out a raft of propaganda designed to discourage people from turning out for Clinton. It distributed anti-Hillary videos, used Facebook to raise nearly $300,000 in campaign donations, and paid Facebook for help in online messaging that highlighted Clinton's support for NAFTA, and played to the "Bernie-Bro"–type misogyny that appealed to some Rust Belt white males who knew they'd been sold out by the Republican Party, but also felt betrayed by the corporatist wing of the Democratic Party, and thus were open to hateful messages about Clinton. If they had a germ of truth in them (while Clinton did shift her trade stance during the campaign, her husband's administration had orchestrated NAFTA), then so much the better.

Some of the propaganda came from the Trump bunker itself. Some was fanned by overseas actors, including Russian agents who wanted to help the reality-TV-star-turned-candidate win the election. Whether or not Trump was actually in hock to Putin (the long-awaited report from special counsel Robert Mueller fell short of declaring collusion, but had plenty of examples of suspicious pre-election communication between Russia and the Trump team, leaving both sides to believe exactly what they had before the Mueller report was in),[7] it was clear that he'd be far easier to manipulate than Clinton, who was hawkish on foreign policy and particularly on Russia.

As Mueller's investigations have shown, the Internet Research

Agency, a Russian firm working for the Kremlin, managed to draw hundreds of users from Facebook and other social media platforms to ads that were designed to damage Clinton. The group's content reached a stunning 150 million Internet users. As Theresa Hong, the Trump campaign's digital director, put it plainly, "Without Facebook, we wouldn't have won."[8]

It was right around that time that Roger McNamee, the Facebook investor and former mentor to Zuckerberg, began noticing strange things online. "I first became seriously concerned about Facebook in February 2016 in the run-up to the first U.S. presidential primary," he told me in 2017, which is around the time he began work on *Zucked,* his book about the topic. As a political junkie, McNamee "was spending a few hours a day reading the news and also spending a fair amount of time on Facebook," he writes. "I noticed a surge on Facebook of disturbing images, shared by friends, that originated on Facebook Groups ostensibly associated with the Bernie Sanders campaign. The images were deeply misogynistic depictions of Hillary Clinton. It was impossible for me to imagine that Bernie's campaign would allow them. More disturbing, the images were spreading virally. Lots of my friends were sharing them. And there were new images every day."[9]

McNamee was becoming increasingly concerned about the effect of social media on election outcomes. He had been shocked, as were most people, by the outcome of Brexit, which defied all public polls in advance of the referendum result. While the British government had initially denied any involvement by nefarious actors using platform tech to influence the referendum, a damning report released by the British Parliament in February 2019 found just the opposite.[10] The report indicated that there was indeed a possibility that the Brexit vote had been influenced by Russian actors, and that at the very least there were big questions about the ongoing role of disinformation in elections, questions that might require a reworking of electoral law, which the committee in charge of the investigation found was "not fit for the purpose" of dealing with companies like Facebook, which "intentionally and knowingly violated both data privacy and anti-competition laws."

As MP Damian Collins put it, "Democracy is at risk from the malicious and relentless targeting of citizens with disinformation and personalized 'dark adverts' from unidentifiable sources, delivered through the major social media platforms we use every day. Much of this is directed from agencies working in foreign countries, including Russia. . . . The big tech companies are failing in the duty of care they owe to their users to act against harmful content, and to respect their data privacy rights."

Collins added, "We believe that in its evidence to the Committee Facebook has often deliberately sought to frustrate our work, by giving incomplete, disingenuous and at times misleading answers to our questions. . . . Even if Mark Zuckerberg doesn't believe he is accountable to the UK Parliament, he is to the billions of Facebook users across the world. Evidence uncovered by my Committee shows he still has questions to answer yet he's continued to duck them, refusing to respond to our invitations directly or sending representatives who don't have the right information. Mark Zuckerberg continually fails to show the levels of leadership and personal responsibility that should be expected from someone who sits at the top of one of the world's biggest companies."

THE SAME CAN certainly be said for Facebook's response to the 2016 U.S. election results. But to anyone who cared to pay attention, the evidence was hiding in plain sight. The more reporting that was done on the role of Facebook and Google and other platforms in the election, the clearer it became that online propaganda had played a role in fanning the flames of issues like immigration, which had been exploited by the pro-Brexit camp during the voting period as well as by the Trump campaign in 2016.

"For the first time," says McNamee, "I realized that Facebook's algorithms might favour incendiary messages over neutral ones."[11] What's more, it was becoming clear that those algorithms might be leading to a more dangerous and polarized world, not to mention a less democratic one.

He decided to do something about it. In October 2016, several days *before* the presidential election, McNamee reached out to Mark Zuckerberg and Sheryl Sandberg. These were people that he counted as friends, and he had every reason to believe they would listen to his concerns. He was, after all, the one who had advised Mark Zuckerberg to turn down Yahoo's $1 billion offer to buy the company early on (a smart move given its current value); he'd also been the one to suggest that Zuck hire Google ad whiz Sheryl Sandberg to be COO of the company.

It turned out that McNamee had been overly optimistic about his relationship with Zuck and Sandberg. Despite his history with the two executives, and his insider status as a longtime Silicon Valley investor, they stonewalled him the same way they had pretty much everyone else over the previous two years.[12]

He wrote later in a *Financial Times* piece, "They politely informed me that what I had seen were isolated events that the company had addressed. After the 2016 US presidential election, I spent three months begging Facebook to recognise the threat to its brand if the problems I observed proved to be the result of flaws in the architecture or business model. I argued that failing to take responsibility might jeopardise the trust on which the business depended."[13] Still no dice.

As McNamee put it to me in interviews in 2017, Sandberg and Zuck had taken a "clean room" attitude toward the emerging scandal. They were going to zip up and protect themselves and the company at all costs, no matter what the ramifications might be. Their model dictated that whatever content resulted in the most clicks, and thus the most revenue, for the platform would win the day, even if that content was designed to manipulate voters or fan the flames of racism and hate. It speaks to how much Facebook—as well as the other major platform firms—has to lose. In 2012, Facebook had a billion users. By 2016 it had double that. Its revenue over that time had more than quadrupled, from $5 billion to $27.6 billion. By the end of 2018 it was $55.8 billion.[14]

In February 2018, the U.S. Department of Justice charged thirteen Russians and three companies with election manipulation, including spreading false and divisive content that helped propel Donald Trump to victory.[15] And as Justice Department investigations have now shown, these entities used U.S. technology platforms, most notably Facebook, which was found to have accepted $100,000 in advertising from Russian actors, but also Instagram (owned by Facebook), Twitter, YouTube (owned by Google), and PayPal to execute their crimes.[16]

And yet, these companies have almost without exception refused to accept responsibility for any of this. A leaked memo written by Facebook VP Andrew Bosworth in 2016 and published by BuzzFeed in 2018 gives a clue as to why: "We connect people. Period. That's why all the work we do in growth [meaning, the privacy-compromising techniques] is justified. All the questionable contact importing practices. All the subtle language that helps people stay searchable by friends. All of the work we do to bring more communication in." In another portion of the memo, Bosworth speculated about what could happen if the connections go bad. "Maybe it costs someone a life by exposing someone to bullies," or "Maybe someone dies in a terrorist attack coordinated on our tools."[17] Apparently, the leadership felt that every possible negative externality was a price worth paying in service of Facebook's higher mission of connecting the world.

In September 2018, journalist Evan Osnos published a telling portrait of Zuck in *The New Yorker*, describing his deny-and-deflect attitude toward not only election manipulation, but the many other scandals that have embroiled Facebook: the breach of user-data agreements, the use of behavioral technology to knowingly manipulate children, the way in which authoritarian regimes like the Burmese junta have used the platform to orchestrate genocide.[18] From the beginning to the end, Zuck's attitude about all these crises was the same—nothing to see here.

"The idea that fake news on Facebook—of which, you know, it's a very small amount of content—influenced the election in any way, I think, is a pretty crazy idea," he said in 2016. And even in the face of all the evidence to the contrary, in the summer of 2018, his stance had not shifted. "I find the notion that people would only vote some way because they were tricked to be almost viscerally offensive," he told Osnos, in a statement that is truly stunning, given the company's development and deployment of technologies that do just that.

These problems did not come without warning. People in the Valley were openly fretting about them as early as 2011. That's when Eli Pariser, the board president of the liberal political organization MoveOn.org, gave a TED Talk about how both Facebook and Google were using algorithms that encouraged people to migrate into political siloes populated only by those who thought as they did. The talk, entitled "Beware Online Filter Bubbles,"[19] came out the same year that Google was introducing its own social network and vying with Facebook to create ever more detailed—that is to say, more valuable to advertisers—profiles of users' online activity. All of the fears that Page and Brin had articulated in their 1998 paper about nefarious actors who might take advantage of Internet users for their own gain were coming true.

Yet, if changing their business model compromised profits, then it was clear which road the platforms would choose. "It took me a very long time to accept that Zuck and Sheryl had fallen victim to overconfidence," says McNamee.[20] Even as evidence piled up, following the U.S. Senate's election meddling report, that Facebook and other platforms had been used to spread disinformation and suppress votes, "I wanted to believe that [they] would eventually change their approach."

It wasn't to be.

Surveillance Capitalism on Steroids

If election manipulation were the only way in which Big Tech was undermining democracy and civil liberty, it would be bad enough. But it's not. Whether or not you voted in 2016, you are at risk of being targeted by the tools of a growing surveillance state in your daily life.

In the 2002 film *Minority Report,* Tom Cruise played a policeman working in a specialized division in Virginia known as PreCrime that apprehended would-be criminals based on foreknowledge of their crimes provided by psychics. The mass surveillance and technology depicted in the movie—location-based personalized advertising, facial recognition, newspapers that updated themselves—are ubiquitous today. The only thing director Steven Spielberg got wrong was the need for psychics. Instead, law enforcement can turn to data and technologies provided by Google, Facebook, Amazon, and the intelligence group Palantir, who have become such big users of data tools that the reality of data-driven crime fighting in the United States has come to mirror dystopian science fiction.

Facebook ad tools, for example, have been used to gather data on people who expressed an interest in Black Lives Matter, data of the sort that—as the ACLU has exposed—was then sold to police departments via a third-party data-monitoring and sales company called Geofeedia.[21] This is far from unusual; the collection and sale of data from not only Big Tech firms but myriad other companies via third-party data brokers is a common practice—indeed, it's the fastest growing part of the U.S. economy.[22]

A 2019 report from the Democratic strategy group Future Majority found that "thousands of companies gather personal information that their customers or clients provide in the course of doing business with them, and then sell their information to large data brokers, such as credit bureaus.[23] In turn, those data brokers analyze, package and resell the information, often as personal profiles. Their customers range from employers involved in hiring and com-

panies planning marketing campaigns, to banks and mortgage lenders, colleges and universities, political campaigns and charities."
And, it should be noted, public entities like law enforcement and
other government agencies.

"In addition, credit card companies and healthcare data firms
also routinely gather, analyze and profit from the personal information of their users." Many of us might be surprised to learn this,
considering the HIPAA restrictions on the sharing of healthcare
data, and the fact that it's difficult for even credit card users themselves to get access to their own credit scores and data. But, as the
report notes, "those restrictions apply only to certain types of financial information—for example, personal bank balances, but not
loan repayment data—and the requirements on healthcare information apply to healthcare providers but not to pharmacies or medical
device producers." That means that the online pharmacy owned by,
say, Amazon, isn't bound by such rules—nor is Fitbit, or any number of fitness apps that might be tracking your health information
and physical movements.

All of that "personal financial and health-related information
can be gathered, analyzed and sold in anonymized forms, which
algorithms can match to most people or simply generate detailed
financial and health-related profiles based on the extensive information that internet platforms and data brokers have on everyone. Finally, the personal data now routinely used for commercial ends are
not limited to the information that people reveal through their activities on internet platforms or through the goods and services they
purchase. In addition, the Internet of Things has projected personal
data gathering into many other aspects of people's lives. For example, smart TVs collect, analyze and sell personal information on
who owns them and what they watch. Smart cars and smartphones
collect, analyze and sell personal information on who owns them
and every place they go. Smart beds and smart fitness bands
collect, analyze and sell information on who uses those products
and their temperatures, heart rates and respiration. Further, the new

generation of wifi-based home devices that respond to people's voice commands—led by Amazon's Alexa, Echo, and Dot, and Google Nest and Google Home—can capture not only personal information about the people who buy and install them but what they say in the range of those devices."[24]

In short, the American surveillance state isn't science fiction—it's already here.

The fact that Silicon Valley companies portray themselves as uber-liberal and praise groups like Black Lives Matter while also monetizing their surveillance is a rich and dark irony, but by no means the only one. Consider Amazon's Orwellian-sounding Rekognition image processing system, which the ACLU recently called upon Jeff Bezos to stop selling to law enforcement officials, saying it was "primed for abuse in the hands of government." The group argued that the system posed a particularly "grave threat to communities, including people of color and immigrants," in a nod to studies that have shown that facial recognition software regularly misidentifies people of color.[25]

But consider also that big data policing was first pushed in the United States several years ago in part as a response to racism and bias. Computation models in policing have been used for some time now; since 1994, a system known as CompStat, which linked crime and enforcement statistics, has been used by law enforcement officials in New York and then elsewhere. The attacks of 9/11 led to a new push for "intelligence-led policing," which connected local and federal law enforcement agencies and their data.

William Bratton, who had run the New York City police force, moved in 2002 to Los Angeles and brought with him "predictive policing" that aimed to use as much data, from as many sources as possible, to predict where crime might occur before it did. Data from myriad sources—crime reports, but also traffic monitoring, calls for public services, and information from the cameras now located all over L.A. and other major cities—could be used to build profiles on individuals. Police could then tag these profiles on a kind of RSS feed, which would give them information about what they were doing, in real time. The result? You might have a traffic violation one day, and depending on what the algorithm knew about you—where you went, what you did—you could be on a police watch list the next day.

The idea was that policing by algorithm would help circumvent human cognitive biases, such as the conflation of blackness and criminality. And yet, the algorithms came with their own problems. Sarah Brayne, an academic at the University of Texas, has studied the use of big data within the Los Angeles Police Department, which has worked with Palantir (which helps collect and organize much of the data) to build predictive models of where crime might occur.[26] She found that big data had fundamentally changed the nature of policing, making it less about reacting to crime and more about prediction and mass surveillance. The bottom line was that the merging of multiple data sources into Palantir's models (just think of all the bits of data about yourself that we've already learned can be collected, collated, and sold, and then add in whatever the police authorities collected themselves) meant that people who had never had any encounters with police might very well end up under surveillance—something that is uncomfortably at odds with the principle of "innocent until proven guilty." What started out as a way to make crime fighting more fair turned out to make it just the opposite.

As Brayne put it in her paper, "This research highlights how data-driven surveillance practices may be implicated in the reproduction of inequality in at least three ways: by deepening the surveillance of individuals already under suspicion; widening the criminal justice dragnet unequally; and leading people to avoid 'surveilling' institutions that are fundamental to social integration." She also noted, importantly, that "mathematized police practices serve to place individuals already under suspicion under new and deeper forms of surveillance, *while appearing to be objective,* or, in the words of one captain, 'just math.' "

Despite the stated intent of the system to avoid bias in police practices, it hides both intentional and unintentional bias in policing and creates a self-perpetuating cycle: Individuals under heightened surveillance have a greater likelihood of being stopped. Such practices work against individuals already in the criminal justice system, while obscuring the role of enforcement in shaping risk profiles that might actually trap them in the system. Moreover, individuals living in low-income, minority areas have a higher probability of their "risk" being quantified than those in more advantaged neighborhoods where the police are not conducting such surveillance.

This is a big deal, not just because it's racist, but also because it misses loads of nefarious activity; indeed, the two issues are tied together. The well-dressed insider trader sitting in his Upper East Side apartment may not trigger the surveillance algorithms, but his crime is far more important and more costly to society than that of a hoodie-wearing minor offender who's had a traffic violation and is now suddenly trapped in a snowballing loop of surveillance policing. This type of social control, of course, "has consequences that reach beyond individuals." The topic of "algoracism" is now a hot one, as activists and civil rights attorneys struggle to stay ahead of the way in which Big Tech has turned policing upside down, with potentially grave civil liberty implications for entire communities.

As alarming as such shifts are, we are only at the beginning of the creation of a world in which everything we do and say, online and offline, can be watched, and used, both by Big Tech and the public sector itself. Consider the Alphabet Sidewalk Labs project in Toronto. The Google parent's "urban innovation" arm, known as Sidewalk Labs, which works with local governments to place sensors and other technologies around cities (ostensibly to improve city services, but also, of course, to garner data for Google), is working on creating a "smart city" in the Canadian city. The high-tech neighborhood, which is being created from scratch along twelve acres of the city's waterfront, will have sensors to detect noise and pollution, as well as heated driveways for smart cars. Robots will deliver mail through underground corridors, and all materials used in the city will be green.[27]

Whether you find such an idea intriguing or creepy, the planning of the entire project has been opaque. Neither the city nor Google released all the details of the project immediately; rather, they've been leaked by investigative journalists. A February 2019 *Toronto Star* piece revealed that the plans for the smart city were much broader than the public had first thought: Google was actually planning to build its own mass transit line to the area, in exchange for a share of the property taxes, development fees, and increased land value that would ordinarily go into the city coffers.[28] Think about that for a minute: One of the richest companies in the world is asking a city government, the sort of entity that it regularly petitions for better infrastructure, education, and services, to give up the money that would help it provide exactly that.

Then there's the question of who gets to keep the data. Sidewalk sensors would be able to track individuals everywhere they go— sitting on park benches, walking across the street, spending time with family members or lovers. Google has pledged to keep the data from all this "anonymous," meaning not associated with any particular individual, and to put at least some of it into a data bank to be used to improve traffic flows and city services. But they haven't pledged to keep it local—meaning that it could be used by Google in any of their operations.

No wonder local protesters are now up in arms about the project as more details emerge. A sprawling "feedback wall" at the site offers visitors a chance to give answers to pre-written questions, such as "I'm not excited about . . ." One visitor had written "Surveillance state." Another scrawled "Making Toronto Great Again."[29] Given the growing outrage, it will be interesting to see if Sidewalk Labs meets the same fate as Amazon HQ2.

But if you think that's creepy, consider Google's Dragonfly search engine project. In August 2018, the Intercept, an investigative journalism website, reported that Google was considering working on a censored version of its search engine for China called Dragonfly.[30] This came as an enormous shock not only to the general public, but to the vast majority of Google's own workforce. The idea of helping the Chinese Communist Party keep unwanted information from its own people, and allowing the party to track any search results back to individuals via their phone numbers, seemed like the antithesis of "Don't be evil."

This was particularly true given Google's history with the Middle Kingdom. The company had been in China once before, when Google.cn launched in 2006. Even back then, Google search was not allowed to guide Chinese people to certain information that the government deemed harmful, like, for example, the student-led Tiananmen Square protests of 1989 that led to the party opening fire on its own people and killing more than ten thousand.[31] But the company decided that simply being in the country and helping such a huge population gain access to some search would help push the government toward greater openness.

It was, in retrospect, a naïve assumption. In China, the only power is the Communist Party. And as is always the case when new technology comes into the Middle Kingdom, the party studied it, controlled it, and eventually bested it, by supporting a homegrown version of Google, called Baidu, which was given more freedom and ability to operate within the country, in exchange for more government control. By 2009, Google had only a third of the search market to Baidu's 58 percent.[32] A year later, Google decided to pull out of the Chinese market, after a hacking episode known as Operation Aurora, in which entities within China targeted Google's intellectual property, its Gmail accounts, and, most important, the identities of human rights activists who had been using the platform. The idea of an autocratic government using the platform to spy on and potentially persecute activists finally forced the company out of China.

China has hardened politically since then, and is today, under the regime of Xi Jinping, arguably as repressive as it's been since the Mao era. The Dragonfly revelations, which the company at first denied and then tried to mitigate, made Washington furious, especially since the news broke around the same time that Google was leaving empty seats at Senate hearings about privacy and antitrust, and was also refusing to work with Pentagon officials on American artificial intelligence projects. U.S. vice president Mike Pence said that the project would "strengthen Communist Party censorship and compromise the privacy of Chinese customers."

Democratic senator Mark Warner and Republican Marco Rubio led a bipartisan group of senators in writing a letter on August 3, 2018, that expressed grave concerns about the human rights and security implications of the project.

"Chinese authorities . . . censor a broad range of news and social media topics," the senators noted, citing "a significant vaccine scandal in China, which may have affected the health of hundreds of thousands of Chinese children," which had "run afoul of censors. News reports indicate that, as of last Monday, the Chinese word for 'vaccine' was one of the most restricted on Weibo, China's Twitter-like microblog platform."

The senators cited *Financial Times* reporting on how authorities were using technology to increase surveillance of the population. Tech companies in China—including Alibaba, Tencent, Baidu, and JD.com—are "inextricably linked with the Chinese state and its security apparatus, and the authorities retain the upper hand in the relationship." The senators concluded that "Google's reported activity to build applications compatible with Chinese censorship demands is all the more concerning in light of relationships that Google has made with these companies, including a technology cross-licensing joint venture with Tencent and an investment of $550 million in JD.com." The letter ended with a simple but damning question: How could "this reported development [Dragonfly] . . . be reconciled with Google's unofficial motto: Don't Be Evil"?[33]

It was a question that Google's own staff began to ask shortly thereafter. How could a company that claimed it could not (or would not) police its platform in the United States, or in many other parts of the world, be willing to work with an autocratic state with no assumptions of privacy or the sort of civil rights enjoyed in much of the developed world?

The answer, unsurprisingly, has to do with business. As many Googlers have told me, China is considered the world's petri dish for digital technology. Even as it's become more repressive, it's become more tech saturated. China has the most Internet users, many of the most innovative services, and some of the world's richest and most powerful companies—Baidu, Alibaba, Tencent, and others—that are completely unhindered by the privacy and antitrust conversations taking place in the United States. Chinese young people are even more digitally connected than their Western peers. This was a market Google was ready to risk everything to be in, as evidenced by a leaked speech from Ben Gomes, Google's head of search. "China is arguably the most interesting market in the world today," Gomes said. Google didn't want to be there just to make money, but "to understand what is happening there in order to inspire us." As he put it, "China will teach us things that we don't know."

Despite tremendous political pressure from politicians, security hawks, and human rights advocates, only when the company's own engineers began arguing for Google to stop working on a censored search engine for China did the company really sit up and listen. It wasn't an easy fight. Former Google engineer Jack Poulson spent more than a month arguing internally for the company to clarify its ethical lines around the project. In April 2019, he wrote an opinion piece for *The New York Times* in which he said he was told by a Google staffer during his exit interview, "We can forgive your politics and focus on your technical contributions as long as you don't do something unforgivable, like speaking to the press." Poulson ignored the advice.[34]

He's since founded Tech Inquiry, a nonprofit that aims to hold Big Tech accountable to human rights standards and democratic norms. He is one of a growing number of technologists who believe that "the time has passed when tech companies can simply build tools, write algorithms and amass data without regard to who uses the technology and for what purpose." Fourteen hundred of Google's own employees signed a letter protesting the opacity of the company's decision to launch Dragonfly. This followed on the heels of a similar employee protest a few months prior about the Pentagon's use of Google AI technology. (The company subsequently ended its Department of Defense contract, a rich irony given that it went ahead with the Chinese work, which it has since also ended.)

The internal crisis over Google's decision to reenter China reminds me of a telling quote from Ken Auletta's book *Googled: The End of the World as We Know It.* This book was written way back in 2009, and the quote was from Columbia professor and tech expert Tim Wu, who has since written his own terrific book on the cognitive problems of Big Tech, *The Attention Merchants.*

As Wu put it back then, "Google is a precocious company. Great grades. Perfect IPO. A typical high school standout. The basic problem is whether they remain true to their founding philosophy. I don't just mean 'Don't be evil.' Will they stay focused on search, on their founding philosophy, which is really an engineer's aesthetic of getting you to what you want as fast as you can and then getting out of the way?" Or, he asked, will Google become "a source of content, a platform, a destination that seeks to keep people in a walled Google garden? I predict that Google will wind up at war with itself."

How prescient. But it's not only Google that is at war with itself. The Internet has become a new kind of battleground for the world's great powers. No longer a single entity, the Internet is becoming a "splinternet" as the United States and China fight to control the way in which it will be run and regulated, as part of a larger great game to control the high-growth, high-tech industries of the future. Both countries are becoming increasingly nationalistic, supporting their own homegrown giants in an effort to win a tech trade war that has become a new cold war. It's a battle that threatens to worsen all the problems that this book has already laid out.

CHAPTER 13

A New World War

In 2018, I experienced something quite odd during a day of reporting on the growing U.S.-China trade and technology conflict. I was talking to a variety of politicians and advisers in Washington to get a better sense of how various political factions felt it should be dealt with. I spoke to, among others, a former Bernie Sanders staffer who was now advising a new crop of progressives like Congresswoman Alexandria Ocasio-Cortez. I also spoke to a high-level Defense Department official. Their views—that China was indeed an existential threat, and that America needed to ring-fence some of its supply chain and prepare for what might be a long-term tech and trade war—were oddly similar, given the divergent political factions that they represented.[1] Both sources recommended the same book to me: *Freedom's Forge*, which lays out the way in which the U.S. auto industry helped the country during the Second World War.[2] The industry, including not just big automakers but also their suppliers, had been marshaled by government officials to ramp up production and aid in the war effort, creating synergies across supply chains that were later leveraged to great effect in the postwar period, when American industry was dominant relative to Europe and Asia—for at least a couple of decades.

The book had been on the reading list at a June 2018 event

sponsored by the National Defense University that brought together military and civilian leaders to discuss the big challenges of the day. Dozens of experts, government officials, and business leaders gathered to talk about the decline in the postwar order, the rise of China, and how the United States could strengthen its manufacturing and defense industries. The goal would be to create resilient supply chains that could withstand not just a trade war, but an actual war.

Amid the broad and varied discussion, speakers shared a general sense that the conventional approach to globalized business, and in particular the idea that American multinationals could do business wherever they liked and however they liked, was over, and that there would be serious ramifications for U.S. industry. "If you accept as your starting point that we are in a great power struggle [with China and Russia], then you have to think about securing the innovation base, making viable the industrial base, and scaling it all," said Major General John Jansen, the event's organizer.[3]

Much of the discussion focused on the way in which America's high-tech supply chain had been outsourced for years to China. Back in the 1980s and '90s, our government made a decision to prioritize service jobs, which were perceived as higher up the economic food chain than manufacturing, and to allow companies to outsource large parts of the industrial ecosystem abroad. At the same time, changes in the financial system, the weakening of unions, and a variety of regulatory tweaks that favored capital over labor led to a short-term, balance-sheet-oriented mentality in which quarterly stock prices (always buoyed by cost cutting) took priority over long-term risks. Consumers and companies were the priority; not citizens, or workers.

This topic, which I covered at length in my previous book *Makers and Takers*, created a bifurcated economy in the United States— one in which there are plenty of software millionaires and burger flippers, but not enough middle-class jobs in between. From a corporate perspective, it has also left many U.S. firms unable to find the products they need outside of China, including key chemical components of crucial weaponry.[4]

That's not only an economic problem—it's increasingly perceived as a security risk. A 2018 Pentagon report to the White House found that the past four decades of manufacturing outsourcing, combined with China's industrial policies and the degradation of STEM and trade skills in the United States, has left America's supply chains—along with its companies, consumers, and citizens—in a vulnerable position.[5] According to Defense Department research,[6] the fragilities are myriad: consumer product security; sagging U.S. innovation in key areas like artificial intelligence; 5G wireless technology; and the fact that China is the sole source for a variety of manufacturing inputs, including many that directly touch the military and defense industries. Most interesting were the industries covered in the report, which included not just the military-industrial complex, but the broader manufacturing supply chain: electronics, machine tools, software, and so on.

While America was once the dominating force in the high-tech sector, over the past few decades of globalization and outsourcing, the landscape has shifted. China is now the world's major supplier of telecommunications equipment. It has moved ahead in mobile apps and payment systems, and it even owns the highest number of patents necessary for the rollout of the new high-speed 5G mobile services that will be necessary to connect the Internet of things and create new growth opportunities for all sorts of businesses. China has also made great strides in artificial intelligence. While companies like Google are still ahead in terms of sheer know-how, far more money is flowing into China's AI efforts; according to a report by the global strategy and research company 13D, roughly 48 percent of global investment in AI is currently going to China, versus 38 percent to the United States.

That's a key reason why, over the past few years, the Pentagon, the Treasury Department, and the Office of the U.S. Trade Representative have been stressing the strategic threat posed to the country by China, which is now officially considered to be America's biggest mid- to long-term adversary. Trump's China troubles are merely the tip of a massive iceberg—a new and most-likely ongoing tech and trade war against a country with a fundamentally different economic and political system. As Daniel Coats, the former U.S. director of national intelligence, put it in a report released in early 2019, "Chinese leaders will increasingly seek to assert China's model of authoritarian capitalism as an alternative—and implicitly superior—development path abroad, exacerbating great power competition that could threaten international support for democracy, human rights and the rule of law."[7] It's a statement that resonates not just among Trump supporters, or conservatives in general, but also in Democratic circles. Increasingly, both sides of the aisle view a clash of cultures between the United States and the Middle Kingdom as inevitable.

This underscores an uncomfortable truth, one that will have major impacts for the private sector. America's system of free market capitalism has for decades allowed companies to do business wherever they like, outsourcing labor wherever it was cheapest to do so, and offshoring profits to the most favorable tax regime. American companies created and led the process of globalization. But in a world in which the United States is now being challenged by China—a country that has its own system and doesn't play by the usual rules of democratic capitalism—there is a growing group of people who believe that it will be crucial for American security and economic interests that there be an untangling of the investment and supply chain links between the United States and China.

This view is often dismissed as China bashing, and certainly there's plenty of that, particularly in advance of presidential elections. But it's also not quite that simple. It's becoming clear that the United States, Europe, and China have different views about what the rules of the new digital economy should be. That means globalization as we know it may shift. In the new digital world, different regions may choose different Internet governance regimes based on different values and priorities. China, for example, may choose to completely disregard user privacy in favor of moving ahead with data collection for the purposes of artificial intelligence innovation. Europe may create public data banks that companies will have to ask to dip into. The United States has so far been lax about regulating Big Tech, but that could change, particularly given that every important 2020 Democratic candidate has made antitrust a core issue.

What will be the result of all this for you and me? There's already a "splinternet," with information and data rights that differ quite widely from one country to another. Companies and governments can do vastly different things with your personal data depending on which country you live in, and you as a consumer and a citizen have different kinds of recourse based on nationality as well. No one country has yet found the perfect system (though I tend to favor the policies supported by both the European Union and California, which start with the notion that individuals own their own data and should have a clear understanding of how it's being used). Choosing the right balance between innovation, competition, and privacy will be crucial to creating both jobs and a stable society.

What's more, it's likely that the geopolitical battles of the future will be waged in cyberspace. Eric Chewning, deputy assistant secretary of defense for industrial policy at the Pentagon, told me, "The US is strategically repositioning for [a] period of interstate competition [meaning, between the West and China]. A competition that recognizes the inter-relationship between economic security and national security." That means that the era of American companies being able to fly thirty-five thousand feet over local concerns may well be coming to an end. Global business itself will change, and with it the nature of our economy and our politics.

There are no clear rules yet for the new era. Should Western companies be allowed to do business as usual with China? And if not, why not? Politicians have different answers to that question, but companies themselves are also taking different approaches depending on their own strategic interests. Amazon, for example, has almost no access to the Chinese market, since Beijing has chosen to support its own homegrown e-tail giant, Alibaba. That means that Amazon has put its resources into becoming the U.S. government's main supply and procurement platform, in ways that have upset some competitors who feel that Amazon has been given unfair advantages.[8] Facebook, on the other hand, is eager to become a greater presence in China—it has been known to keep data locally in China, and share it with app companies there, something that could not only compromise the privacy of U.S. citizens, but present all sorts of hacking risks, à la Google's experience with Operation Aurora. Even Apple, which has tried to brand itself as a defender of privacy, has capitulated in China. While the company has been quite protective of user data in the United States—it refused to help the FBI break into a locked iPhone during investigations of the 2015 San Bernardino terrorist attack—it plays by local rules in China. Beijing has, for example, forced the company to move all of its iCloud data centers for Chinese customers to the mainland where they will be run by a local company that does not need to comply with U.S. laws about data protection. So much for privacy, at least for Apple users in China.

Google, for its part, does not appear to share user data in China the same way that Facebook does. (PR reps insist that they don't hold any data locally in China, but rather keep it in cloud servers, supposedly away from the eyes of the government.) But it has operated a research facility in Beijing and explored the idea of launching a censored Chinese market search engine, all while working with the U.S. Department of Defense on various projects. It's hard to imagine a U.S. defense contractor like, say, Raytheon, *ever* having been allowed to have the same sort of arrangement with both the United States and China. And yet, Big Tech companies that do the most cutting-edge research in areas such as artificial intelligence are arguably more important to the Defense Department than the old military-industrial complex represented by Raytheon. And they still get to play both sides of the field.

All of this is top of mind for the Pentagon, which has begun having ongoing conversations with many U.S. multinationals about how they do business with China. "Based on our conversations, corporations recognize that the strategic conditions are changing," says Chewning. "As China continues to pursue the objectives laid out in Made in China 2025 and actively distorts the economic playing field in favor of their national champions, Western businesses will also likely need to re-evaluate their confidence in the long-term China business case." Already, many companies are doing just that, rethinking supply chains that run through the South China Sea, moving production out of China to places like Vietnam and Mexico, and considering how political risk and conflict between the two powers may affect their businesses.

Translation: We are likely at the biggest geopolitical turning point since the Second World War. Big Tech will be at the center of it. And the Internet itself will be the battlefield. It's a far cry from the happy, connected world that the Big Tech utopians once dreamed about.

The Rise of Techno-Nationalism

Why is this happening? In part, because American and European policy makers made the wrong bet about China. The conventional wisdom has always been that as China became more developed, it would also become freer. But after the financial crisis of 2008, the cracks and hypocrisies inherent in America's own system were exposed, and the Chinese became understandably worried about how opening up—financially, economically, and politically—could put them at risk and make them vulnerable to outside forces (like, for example, rapacious Wall Street bankers who could bring down global financial systems).

The result was that the economic opening and privatization of the previous decades began to tail off. The Chinese began to assert themselves, and their own model of state control, on the global stage. No longer biding their time and hiding their brilliance, as the old Deng Xiaoping quote goes, a new group of technocrats led by Xi Jinping began to consolidate power and export their own political values, capital, and technology to other countries.

Some of these moves, like new investments in Africa, and the One Belt, One Road strategy, which aims to connect China through the old Silk Road all the way to Europe, creating new economic and political alliances in the process, have been commended, though not by everyone—the idea that China will somehow be able to create fair and productive new alliances in difficult parts of the world where others have failed seems naïve at best. To be fair, the jury is still out on whether China's economic diplomacy efforts, which are looping in many European countries as well, will be a good thing for both China and the world at large. On the one hand, it's fair and right that China should play a greater role on the global stage, as the world's number two economy. On the other, how to do business with a state-run autocracy in which individuals have absolutely no right to privacy is a massive conundrum for Western companies and countries.

But in some ways, for China it really doesn't matter. That's because China is a country that looks very much like the United States in the post-WWII period—a large, single-nation growth market with tremendous potential gone that will pull many other countries into its orbit. The Chinese have spent the past several years moving up the value chain, and now they want to use their system of state capitalism to give preference to homegrown players, many of whom are just as competitive as American firms in the home market. Indeed, it's a wonder to me that American firms ever believed they'd be able to compete with Chinese firms in their home market on a completely equal footing.

I think frequently of a trip that I took to China, right around the time of the Edward Snowden NSA leaks. I met with a People's Liberation Army general (a woman, interestingly) and asked her about Chinese state-sponsored intellectual property theft and the notion that technologies taken from the West might be used for both economic and national security advantage. She made it quite clear that "capitalism with Chinese characteristics" meant that there was no real boundary between corporate interests and national interests. Indeed, she seemed to think it a bit naïve that anyone would assume otherwise. It was the country that mattered most, not the company.

Chinese venture capitalist and AI expert Kai-Fu Lee (the man who first launched Google into China in 2006) told me in 2018 that he believes that the United States and China will continue to move in different directions, each developing their own separate technology ecosystems, with the big race being between Google and China, in terms of who is able to develop the most sophisticated artificial intelligence systems (AI being the most strategic technology of the future). Lee believed that the growing trade and tech war could cut off capital flows and trade in technology between the two countries. But, he said, that was all right with him. China, he believed, would be more than capable of developing a prosperous technological ecosystem of its own. Pointing to already dominant Chinese brands like Xiaomi, which now sells more mobile phones in China than Apple, he said, "The question in the future will be, 'Why buy a Western brand?' I think you'll see China owning the digital ecosystem not only at home, but also in ASEAN [Association of Southeast Asian Nations] and many Middle Eastern countries."

It's easy to make that argument. There are vastly more Internet users in China—some 800 million—than in America. The country has become a largely cashless society in which the majority of the population use locally developed apps for everything from mobile banking to food delivery to bicycle rental. As big as American Big Tech firms are, the Chinese giants Alibaba, Tencent, and Baidu are even bigger in relative terms. The Chinese, who've come of age in a system with no assumptions of Western-style personal freedom, seem happy to give up personal privacy in exchange for the many conveniences of Big Data. They will, for example, consent to medical sensors implanted in bodies to monitor health, and "social scorecards" in which citizens are given up or down marks for nearly every move they make, a science fiction–like system that results in easy loans and better housing for those with "top marks" in the country's big data scoring system, but also discrimination and the inability to get a job for those that have low marks.[9] If you think this sounds very much like an episode of the dystopian TV drama *Black Mirror,* you're right—except in China, the system is real.

But there are massive and obvious downsides within the Chinese system, as well. Citizens can have their social media accounts closed down for the smallest infractions. They can be jailed if their scorecards have the wrong marks. Their every move can be tracked via facial recognition scanning systems—in schools, hospitals, and even in their homes. The result may be an advantage in Big Data. It may also be a return to a Maoist era of total state control and repression.

"There is no longer any freedom of speech in China," says Jia Jia, a Chinese blogger who has written, at great personal risk, about the rise of the surveillance state in China. "In the end, no one will be spared."[10] What's more, the Chinese have sold some of these repressive technologies to other countries, as part of a new effort to reshape geopolitics and extend their influence. A 2018 report on the rise of digital authoritarianism published by Freedom House[11] found that Beijing had exported its surveillance technology to at least eighteen other countries, making it easier for Zambia, Vietnam, and other governments to crack down on their citizens.[12]

Top-Down or Bottom-Up?

Few Americans or Western Europeans would say that this state of Internet governance is desirable. But if you put aside the human rights implications of China's top-down surveillance state, there is still an important question to be asked: Is digital innovation best suited to an environment of decentralization, in which many firms in the private sector working under smart regulation and within a truly free marketplace are allowed to compete? Or is the best model for the future one of centralization, in which a top-down surveillance state can collect all the data it wants, and allow the companies that it has handpicked to do with it what they like?

China is obviously betting on the latter. Xi has, during his tenure, tightened state control of the market and made it tougher for a host of companies—from Qualcomm to Apple to Visa and Mastercard—to do business in China. Beijing is also exerting more control over the high-growth technology sector, requiring both foreign and domestic companies to engage in information censorship and cooperate with state security efforts. In an age of artificial intelligence and big data, the story goes, China will have an advantage over the United States because there is no civil liberty debate to get in the way of the surveillance state. With unfettered access to all the information generated by the world's largest population, the Chinese tech sector will move ahead quickly.[13]

It has been assumed that 5G, the fifth-generation mobile technology that has yet to become a large-scale commercial reality, will be dominated by China. The ongoing U.S.-China trade war makes it difficult to see how this will play out. China's own homegrown 5G chip maker, Huawei, could be severely handicapped by Trump administration attempts to put restrictions on Huawei doing business in the United States and with American firms. Still, proponents argue that China is moving ahead quickly to build out the necessary infrastructure, while the United States and particularly Europe lag behind. While Huawei equipment, which is relatively cheap, is being adopted by many countries, the U.S. chip champion, Qualcomm, was until mid-2019 bogged down in a multiyear, multicontinent legal battle with Apple that drew resources and attention away from its own rollout of 5G.

This underscores one of the great ironies of the current U.S.-China trade and tech battle. The United States wants to kneecap

Huawei (a company founded by a former PLA general), by pushing allies in Europe and other countries to not use its equipment, and by preventing U.S. companies from doing business with the firm. This is in many ways understandable, given widely reported revelations of the extent of Chinese IP theft and cyber-espionage.[14]

But Qualcomm hasn't been hurt so much by Huawei as it has by the monopoly power of Apple, which, when it became big and powerful enough, simply decided it didn't want to pay Qualcomm's patent license fees, and held the company up in court in three continents for many years. That's a point worth remembering: Part of the tech and trade war is about China playing by its own rules. But another part is about the United States allowing the largest Big Tech companies like Apple to become so large and powerful that they can set their own terms in the marketplace in a way that may be economically and politically disruptive to the country at large.

Still, even if the United States were moving faster on 5G, many believe that it will be easier for a surveillance state such as China to own and harness the data that will be transferred via the 5G chips that will exist in all sorts of products from tires to tennis shoes to fetal heart monitors. That would, in turn, allow Beijing to harness the productivity benefit from such data more quickly. The key idea behind this thinking is that we've left the "innovation" stage of artificial intelligence use, and the only thing that matters is the data—whoever can get the most of it, wins.

In this line of thinking, there are no more great leaps forward to be made in AI innovation—it's all about who can create the biggest surveillance state. Beyond that, China proponents argue that the Communist Party has the advantage of being able to direct the resources of such firms toward its own industrial policy aims, pushing companies like Alibaba to build out rural broadband, for example. The state provides support for strategic industries such as robotics, semiconductors, and electric cars.

All of these things may prove to be competitive advantages. But what we also know for sure is this: The companies that have best commercialized the Internet—not just American giants Google, Facebook, Amazon, and Microsoft, but also the Chinese leaders Baidu, Alibaba, and Tencent—did so when they were young and had more decentralized cultures. There is plenty of academic evidence to show that breakthrough innovations are more likely to come from lone academics than corporate (or state) behemoths.[15] If you believe China will own AI, as Kai-Fu Lee does, you also have to believe that we've left the era of innovation and that data-driven surveillance states run by large countries and companies are the only path forward. That is, I believe, too big a leap to make.[16] One of the few things that we can say for certain about innovation is that its path is rarely what we predict that it will be.

While China's centralized control may be a short-term advantage, you can question whether it will pay off in the long term. "In the next three to four years, centralization will be beneficial, but in five to ten years, you may get the problems of the brittleness of centralized control," says Arthur Kroeber, the managing director of Gavekal Dragonomics, an influential consultancy and research group focused on China. Think of the disasters of central planning under Mao, or even the recent inability of China to dominate the automotive industry.

As the historian Niall Ferguson points out in his book *The Square and the Tower*, rigid, hierarchical structures tend to be undermined by disruptive technologies. Authoritarian capitalism and the Internet may not be suited to each other.[17] Nicholas Lardy of the Washington-based Peterson Institute for International Economics believes that top-down control and the increasing drag of state companies in China have actually resulted in a massive productivity slowdown since the global financial crisis. Reversing this will require less state control, not more. After China opened up in the 1980s, it attracted more than $1.7 trillion of foreign investment and made a huge contribution to global growth by allowing the private sector to diversify and grow. Today, as the surveillance state exerts more and more control, both overall capital flows and growth are decreasing.[18]

America's "National Champions"?

Of course, it's possible that China could be at a turning point—much will be revealed in the next few years about which country's digital strategy will work best. But one thing is certain: America will be able to capitalize on its own historical strengths only if U.S. companies are able to compete and innovate on an even playing field, and exploit the advantages of the free market system, which are its decentralization and possibility for creative destruction and innovation from all sorts of companies—big, medium-sized, and small alike. Unfortunately, as we've learned throughout this book, that's less and less the case. Big Tech in both the United States and China has become ever more monopolistic, creating large and relatively closed-off systems that attempt to control "talent and resources on breakthroughs that will remain mostly 'in house,'" as Lee puts it in his book *AI Superpowers*. Silicon Valley leaders, such as Facebook's Mark Zuckerberg, cite this as a virtue. And, when threatened with more regulation, they argue that breaking up tech behemoths would prevent them from competing with the Chinese.

Back in 2018, Zuckerberg testified before Congress about the company's involvement in Russian election manipulation. The most interesting tidbit to emerge from that event, which resembled nothing so much as a four-hour tech support call, was a photo of Zuckerberg's talking points taken by the Associated Press.[19] One bullet point made it clear that, if he was questioned about competition issues, he should argue that a breakup of Facebook would "strengthen Chinese companies." Yet for several years, Facebook allowed the Chinese telecom company Huawei, which has been deemed an official security threat by the U.S. government, and other Chinese groups to access detailed data about users and all of their friends, including their work history, personal relationships. and religious affiliations.[20] Facebook eventually closed down the Huawei partnership. Whether or not it still has other data partnerships going in China is unclear.

The fact that Facebook shares personal data with third-party companies in ways that users are completely unaware of isn't news to anyone at this point. The social network has had multiple data-sharing partnerships with dozens of device makers, including Samsung and Apple, for years now. These deals—in which the company profited from sharing user data even as it was promising users it was protecting it—may well have been in violation of Facebook's 2011 agreement with the Federal Trade Commission, in which it promised not to share users' personal data with outside partners. That's one reason that in 2019, the FTC slapped Facebook with a $5 billion fine, the largest in history for a technology company.[21] The FTC has since launched a new antitrust investigation.

Facebook not only shared data with major Western companies and shadowy entities like Cambridge Analytica; it has also shared user data with Chinese companies that operate within a repressive surveillance state.[22] The privacy and civil liberty risks inherent in doing so are one reason that U.S. authorities have begun cracking down on Chinese businesses working in the United States in sensitive areas. In the spring of 2019, for example, the Chinese firm Kunlun Tech was required to divest itself of the American social network Grindr, on the grounds that data from the LGBTQ dating app could potentially be used by the Chinese state to blackmail people with U.S. security clearances, thus compromising U.S. national interests.

This underscores what may well be the biggest Big Tech hypocrisy of all: the ways in which Facebook, Google, and even Apple do business with an autocratic government for profit, even as they purport to be America's "national champions" in the race against China to control the world's most strategic, high-growth industries.[23] It's hard for me to imagine why, if the U.S. government is prepared to force Chinese firms to divest themselves of sensitive apps, U.S. tech companies that could be compromising data by doing business in the Middle Kingdom aren't coming under even more scrutiny than they already are.

There are some similarities between how the two entities—Big

Tech and the Chinese surveillance state—act. Libertarians like Pay-Pal's Peter Thiel argue that in the world of Big Data, "freedom and democracy are [no longer] compatible," something that Chinese leaders would themselves probably agree with. Other titans of Big Tech have argued that if we are indeed in a race for the future with China, then we shouldn't let freedom get in the way of competition.

In 2017, for example, former Google chair Eric Schmidt gave a keynote address at the Center for a New American Security's Artificial Intelligence and Global Security Summit, in which he detailed his own consulting work around cyber issues for the Department of Defense, and raised the specter of China overtaking the United States in the most cutting-edge technologies, including AI. "By 2020 they will have caught up. By 2025 they will be better than us and by 2030 they will dominate the industries of AI," said Schmidt. "Just stop for a second. That's the [Chinese] government that said that. Weren't we the ones in charge of AI dominance here in our country? Weren't we the ones who invented this stuff? Weren't we the ones that were going to exploit the benefits of all this technology for betterment and American Exceptionalism in our own arrogant view? Trust me, these Chinese people are good."[24]

The message here is that China is about to best the United States, and the only way for America to stay on top is to let the big stay big, and allow Silicon Valley to make the decisions about how to move forward in the digital economy. It's a position that tech titans have argued in the offices of powerful members of Congress, on the speaking circuit, at dinner parties, and in all the places where they exert their influence. It's a line that Google is pushing via academics who write and speak on the topic, and via the think tanks that it funds.

Witness the transformation by the influential Information Technology and Innovation Foundation president, Rob Atkinson, who once championed small and mid-sized firms. Now, he's written a book, along with Michael Lind, entitled *Big Is Beautiful: Debunking the Myth of Small Business*.[25] While it's true that ITIF has taken some positions on different issues that go against the Google line, it's telling that the book seems directed toward debunking the "new Brandeis" theory of antitrust put forward by Tim Wu and Barry Lynn, who was kicked out of his position at the New America Foundation after publicly lauding the European Union's antitrust case against Google.

It's true that the Chinese are making rapid strides in intellectual property, innovation, and economic competitiveness. And yet, the war for the technologies of the future is by no means won. China has been stacking the international telecom-standards bodies with their own representatives over the past few years, and they have a head start on 5G network construction. But that may or may not equate to a lasting technological advantage. China was late to adopt 3G, but that did not stop their progress in 4G, for example. There is no reason to think that the United States, or even Europe, cannot catch up in 5G. The Chinese government does have the advantage of being able to direct state-owned operators to build infrastructure, and its companies don't have to purchase spectrum before offering 5G services. Yet the first country to have an operational 5G network will not be China, but South Korea.

It's important which countries build their networks first, but it's not the only factor in success. The real advantages of 5G are going to come from the way in which individual companies and industries exploit the potential gains from 5G—and it's difficult to know where they will come from just yet. "Just as no one predicted that one of the major uses of 4G would be a new way of calling taxis, the most important uses for 5G technology are also difficult to predict before [they are] actually available," says Dan Wang, an analyst for Gavekal Dragonomics.

Go back even further, and consider how the industrial revolution developed: The big things came first, like electricity and the combustion engine. But they were followed by spates of innovative products and services, ranging from automobiles and domestic appliances to wind turbines. There is no reason to think this time will be different—or that China, the United States, and Europe can't all have a piece of the pie. The battle for 5G isn't set—and it doesn't have to be a zero-sum game.

In the United States, though, shared prosperity will be obtainable only if the government moves quickly to create a more supportive environment for real innovators (coordinating with other countries on 5G standards, for example), and if Big Tech doesn't monopolize the next generation of innovation. We should dismiss the cynical "Better us than China" argument about why Big Tech shouldn't be the subject of greater regulation.

Thinking of Facebook and Google, *both of which do business in China,* as any kind of counterweight to Chinese tech nationalism is completely nonsensical. These companies may know how to monetize U.S.-government-sponsored innovations like the Internet, but they are profit-making companies, not national champions. In fact, in the past, they have frequently argued that they have no responsibility to be the latter—in a free market country like the United States, they can offshore as much cash as they want, and invest where they like. Indeed, the U.S. tech lobby frequently bashes European policy makers for nationalism and preferential treatment for their own tech companies. To use the "It's us against China" argument to prevent further scrutiny or regulation, particularly given the extent to which these technologies are now being used to undermine liberal democracy not just in the United States but elsewhere, goes beyond the hypocritical to the completely cynical.

Building a Better System

Monopoly power, more than China itself, threatens the most important advantage that America has in the race for the high-growth industries of the future: a fair and open market system. In the United States these days, it is not only small firms that are being squashed, but medium-sized and even large players, too. The country cannot compete with China on top-down approaches to competitiveness.

Nor should it. But it could reassert the merits of a more decentralized system by curbing the companies that threaten it.

Doing so will require more than just regulatory and political willpower. It will require a fundamental rethink about how the market system works, both at home and abroad. The government is right to want to look out for national security interests and to have a hand in how strategically important sectors are managed. I agree with defense hawks who believe the United States should rely less on Chinese equipment and rebuild its own industrial infrastructure. But grandstanding and China bashing for political effect, as we've seen President Trump do over the past few years, is not the way to Make America Great Again; fueling the homegrown innovation ecosystem is.

The U.S. government has a terrific record in terms of funding blue-sky research that results in huge economic value for the private sector—touchscreen technology, GPS, and the Internet itself came out of the Pentagon. We should be bolstering rather than cutting funding for such core research, and perhaps even allowing the public sector to take a greater cut of the profits if the research is commercialized, as Nordic countries and Israel do. That would help offset the popular criticism that results when companies such as Apple, Google, and Qualcomm, after benefiting greatly from publicly funded basic research, end up stashing much of their profits offshore.

We should also think deeply about the existential challenge posed by the Chinese state system—not to copy it, but to address the problems that it has illuminated within our own system. The cracks within the neoliberal framework can no longer be plastered over. It's good and right that myriad Democratic candidates for 2020 are thinking through various ways to bring economic globalization and multinational business practices down to earth, and put corporations more in service to citizens and society, rather than allowing them to operate on such a selfish, short-term basis. If capitalism is going to be sustainable, people have to believe it's working for them. (Today, only a minority of millennials, the largest voting group, profess to believe in capitalism.)[26] If there's any lesson to take from China's system, it's that winning in the future will require a long-term focus. In the United States, the answer isn't to shift toward socialist-style state planning by creating national champions, but rather to enrich the innovation ecosystem as a whole, and to reshape the market system to focus more on stakeholder rather than merely shareholder well-being—as a number of Democratic candidates, including Elizabeth Warren, have proposed.

Meanwhile, rather than swallowing the claim that Big Tech companies are somehow national champions, we should take a closer look at our own digital ecosystem. Large U.S. incumbents are crushing innovation. Educational reform is desperately needed to train workers for jobs in which they will not be displaced by robots. America's largest and richest companies are keeping the vast majority of their cash abroad, where it can't help fund the very things that the industry claims it needs from the public sector. We can best bolster growth not by protecting U.S. companies from overseas buyers, but by addressing these concerns and creating a fairer marketplace.

CHAPTER 14

How to Not Be Evil

I f there is a commodity more valuable than data, it must be time. I dream of being able to wake thirty minutes late, or linger over newspapers (yes, the print kind) and coffee for hours without considering a mental to-do list, to look at my email at the end of the day and see nothing there left to attend to, or to have several years—rather than twelve months—to write a book like the one you've just read. Sadly, in our fast-twitch, high-speed, always-on digital world, slowing down is the hardest thing to do.

But if there was ever a moment to take a breath and pause, it's now, as we begin to understand and grapple with the rise of Big Tech, and all the many things—good and bad—that it has wrought. The problem is urgent. But that doesn't mean we should be reactive. Indeed, the enormity of the changes facing us, and what's at stake if we get it wrong, argues for taking the time to contemplate our next moves as a society in a thoughtful way. After all, quickly conceived, insufficiently informed wisdom is a terrible side effect of the Big Tech era. All too often, we come up with some firmly held position after a few glances at our Facebook page or Twitter feed—a position based more on feeling than on fact.

I worry, too, that we may find ourselves in the sort of regulatory paradigm we saw in the financial sector post-2008. Back then, lobbyists and vested interests on both sides of the political aisle came up with a complex stew of new laws. Some were good, some bad. The sheer complexity created plenty of loopholes for corporate lawyers to jump through. While risk-taking was curbed at some individual institutions, the system as a whole was made no safer. In the haze of complex, technocratic debate, we lost sight of the only question that mattered: How can we create a financial sector that serves the real economy?

We need to ask that question now about the new technologies that are proliferating all around us. The structural shift from a tangible to an intangible economy—one that makes the industrial revolution look relatively minor by comparison—should trigger deep thinking about a host of big topics: digital property rights, trade regulations, privacy laws, antitrust rules, liability rules, free speech, the legality of surveillance, the implications of data for economic competitiveness and national security, the impact of the algorithmic disruption of work on labor markets, the ethics of artificial intelligence, and the health and well-being of users of digital technology.

Even taken individually, these are deep and complex issues. But they need to be taken together, because each one impacts the others. This challenge is one that requires policy makers to have robust conversations with experts from a broad range of disciplines about what the new framework for economic growth, political stability, personal liberty, and health and safety in our complex new digital world should be. In the wake of 2008, policy makers trying to fix the financial system were captured by Wall Street itself—the vast majority of the consultations taken on the most contentious post–financial crisis rules were taken with the very people who were being regulated.[1] It wasn't a good look, and it had major political ramifications in the sense that Americans were left feeling that the system was rigged. We need to make sure to not make those mistakes again. As we think about how to harness the power of technology for the public good, rather than the enrichment of a few companies, we must make sure that the leaders of those companies aren't the only ones to have a say in what the rules are.

One step in this process might be to create a national commission on the future of data and digital technology, ideally a bipartisan and independent one, which would lay out all of the issues at stake in a report to Congress. It's true that "blue ribbon" commissions often come in for political criticism—they are too big, too vague, and too slow moving, or so the critique goes. As a former boss of mine once told me, "Nothing ever gets done in a super committee." That's fair. But given the complexity and importance and interconnectivity of the issues at stake, I think a national commission set up not to make rules, but simply to lay out the issues, is a necessary first step. I'm often struck, when in Washington discussing these topics with policy makers, how even the most well-meaning and thoughtful of them tend to focus on only one or two issues involving Big Tech, rather than considering the whole picture. And the companies, of course, would more often than not like to keep it that way, to the extent that a more limited viewpoint on the part of politicians often benefits them.

That's another key reason for a national committee—it would give citizens a sense that these issues are being debated by a *democratically elected body,* rather than a bunch of power brokers in a private room somewhere. Such a body would have a finite timeline for producing a clear and concise report, in plain English, which could then be disseminated for public debate. It's hard for me to imagine how policy makers, let alone the public, could begin to understand all the implications of the digital age without such a road map to start the discussion.

Such a commission would not only lay out the issues, but also consider them in the context of four stakeholder groups: citizens, workers, consumers, and businesses. Then, there should be a vigorous national debate about how to move forward with creating a framework for our next phase of economic, political, and social development. This is, of course, a process that could and should be conducted in any number of countries. How can we make sure that the digital age is one that enhances well-being, creates sustainable growth, and supports rather than erodes our system of liberal democracy? These are the big questions, and we should ask and answer them well.

In the interest of that, rather than presenting a series of pat solutions to complex issues that will have multi-generational impacts, I will spend this chapter putting forward some categories of concern that I believe deserve careful consideration, as well as my own ideas about how we might begin to think about them.

Creating Boundaries for Big Tech

It's important to remember that the rules for capitalism in general are not handed down on stone tablets—we make them, and we can remake them. I believe that both liberal democracy and personal freedom and security are at risk unless we set some boundaries around Big Tech. Below are some thoughts about what the parameters of digital regulation might look like.

For starters, it's worth remembering what we have long known but seem to have forgotten: Industry self-regulation rarely works. From turn-of-the-century railroads, through energy markets in the 1990s, to the financial industry circa 2007, there are many examples that bear this out. The tech industry is only the latest. The contrition and apologies of the Big Tech executives who have sat in front of Congress any number of times since 2016 haven't added up to any significant shift, either in business model or philosophy. Rather, their vague promises to "do better," and spurious claims that they simply can't police activity on their own platforms, just underscore the need for a cohesive regulatory framework around private companies that have amassed too much power.

Smart regulation is, it must be said, very tough to craft. Finance is, again, the perfect example of this: The complexity and global fragmentation of the post-2008 regulatory landscape introduced its own risks into the system, which was one reason it was possible for the Trump administration to justify tearing some of that regulation apart. But that's not a reason to not engage in the process. If there's one thing that's been shown in recent years to be worse than imperfect regulation, it's no regulation. So how do we create a framework for government oversight of Big Tech that protects consumer and societal interests, curbs growth-suppressing monopoly power, and allows us to keep the digital conveniences we depend on?

One way would be to reconsider the legal exemptions that Big Tech enjoys from responsibility for what happens on its platforms. It's a topic that was brought into sharp focus by New Zealand's prime minister, Jacinda Ardern, in the wake of the March 2019 massacre of fifty worshippers in two mosques in Christchurch, New Zealand, which was live streamed in a seventeen-minute video. That video was then uploaded 1.5 million times within twenty-four hours on Facebook and uploaded on YouTube at the rate of one video per second.[2] As she put it in a powerful speech following the episode, "We cannot simply sit back and accept that these platforms just exist and that what is said on them is not the responsibility of the place where they are published. They are the publisher, not just the postman. It cannot be a case of all profit, no responsibility."[3]

The Communications Decency Act section 230 exceptions that allow platforms to get away with the dissemination of hate and violence in a way that no other type of media can are ripe for review. Rethinking them will be tricky: There is a risk that platforms will be overzealous in the policing of hate speech if they are on the hook for it legally, and that could in turn have a chilling effect on free speech in general. But it's clear that the status quo isn't working. Countries such as Germany have passed laws requiring platforms to delete illegal content within twenty-four hours or face large fines. Others, like Australia, are considering similar legislation. While the First Amendment would make any similar law in the United States far narrower, and trade-offs between under- and overregulation of content remain, the fact is that it's time for platforms to admit that they aren't the town square, but advertising businesses that monetize content, just like any other type of media business. It's clearly unfair—not to mention dangerous—for them to act otherwise.

The other major regulatory shift we should consider is a separation of platforms and commerce to create a more equitable and competitive digital landscape. The power of Big Tech is all too reminiscent of the power held by nineteenth-century railroad barons. They, too, dominated their economy and society. And they, too, were able to price gouge, drive competitors out of business, and avoid taxation and regulation, largely by buying off politicians. Yet eventually, they were curbed by a number of regulatory changes, including the creation of the Interstate Commerce Commission, which included some provisions that the industry favored, as well as many it lobbied against. Rather than crush innovation, the ICC ushered in a period of prosperity by allowing technology benefits to be widely shared.

Many experts would argue that Big Tech companies with strong network effects are natural monopolies and should be regulated like utilities, with government oversight to make sure that they can't prevent competitors from using the networks fairly, or use predatory pricing or unreasonable terms of service to gain undue control over the Internet—which is, of course, the railroad of the twenty-first century. This may mean bringing back older antitrust ideas, such as the concept of the essential facilities doctrine, which the Supreme Court used in 1912 to compel the railroads that controlled the only bridges over the Mississippi River in St. Louis to grant access to their rivals on nondiscriminatory terms. It's easy to see the parallels for Google, Amazon, Facebook, and Apple today—all of which hold huge power over their respective ecosystems.[4] This idea is, of course, already in the public conversation, and has been proposed by a number of policy makers, including Massachusetts senator and Democratic presidential candidate Elizabeth Warren. She has also compared Big Tech to the railroads and believes that companies with more than $25 billion in global revenue should not be allowed to own a platform "utility" and also be a participant on that platform.

Finally, we should consider taking antitrust policy back to a broader interpretation of political power, as I described at length in chapter 9, one in which societal welfare, rather than just that of consumers, is taken into account. This is the only way to enforce fairness and economic competitiveness in an era in which the big technology companies, who have blanketed Washington with money and lobbyists, are exerting kudzu-like control over the political economy.

Who Profits from Our Data, and How Can We Better Share the Benefits?

I have no doubt that even if they are well regulated, Big Tech firms will continue to turn disproportionate profits, because as we've already learned, their key inputs—our data—are had for free. In an era in which most wealth will live in data, intellectual property, and other intangible assets, it will be important to come up with more equitable ways to share that pie.

There are those who believe that even to have the discussion of how better to share the spoils of surveillance capitalism is to capitulate to it. You can certainly make that argument. But the fact is that the horse is out of the barn. I'd argue that while we are spending the time to figure out exactly how to regulate and curb the power of Big Tech, we should also make sure that it isn't mining our biggest natural resource for free.

As we learned earlier in this book, the extraction of personal data is America's fastest growing industry, one that will be worth $197.7 billion by 2022 if current trends hold—more than the total value of American agricultural output.[5] If data is the new oil, then the United States is the Saudi Arabia of the digital era. The leading Internet platform companies are the new Aramco and ExxonMobil.

But the tech platform companies are not the only ones in the digital surveillance business. Data brokers such as credit bureaus, healthcare firms, and credit card companies collect and sell all sorts of sensitive personal user data to other businesses and organizations that do not have the scale to collect it themselves. These include retailers, banks, mortgage lenders, colleges, universities, charities, and—as if we could forget—political campaigns.

This is one reason we haven't seen more companies outside Silicon Valley pushing for antitrust action against the big technology companies—they are the ones buying what the Valley is selling. The advent of the Internet of things, in which Web-enabled sensors are embedded in objects all around us, will exponentially expand the opportunities for digital resource extraction. Every company is getting into this business. As a result, we may not be able to simply regulate away all the problems that are being posed by surveillance capitalism.

That's why it is worth considering whether the companies that extract our digital oil should have to pay for it. California has also proposed a "digital dividend" paid by data collectors to the owners of this resource—all of us. It is akin to the way Alaska and countries including Norway have created wealth funds into which a percentage of revenues from commodities are invested for the benefit of future generations. The extractors can afford it. Google and Facebook have high double-digit profit margins because they do not pay for their raw inputs—our data. But we should own our own personal information. And if the extractors use it, they should have to compensate us.

The four major categories of data harvesters—platforms, data brokers, credit cards, and healthcare firms—could pay every American who uses the Internet a set fee, using a portion of their own revenue. Or the extractors could be forced to put a portion of that money into a public fund that invests in education and infrastructure. Education in particular would be an excellent use of such funds, given that all the shifts that I've outlined in this book will require the retraining of a twenty-first-century workforce; it seems only fair that Big Tech—which often complains about the lack of adequate education in the United States—should have to help pay for that. Meanwhile, a 50 percent levy on digital revenue could likewise plug the majority of an American infrastructure spending gap estimated to be $135 billion by 2022.[6] That seems more than a fair exchange for allowing the data collectors free access to the country's most valuable resource. If data is a resource, then perhaps we need a sovereign wealth fund for it.

Taxing data extractors cannot, however, be a get-out-of-jail-free card that allows them to run roughshod over individual privacy or civil liberty. For users of platform technology, transparency could be increased with "opt-in" provisions that allow them more control over how their data is used (as is the case with the EU's General Data Protection Regulation, and the even tougher proposals in California). The "opt-in" language should be clear and simple, with the burden of proof for violations on companies rather than individuals. Big Tech companies should also be required to keep audit logs of the data they feed into their algorithms, and be prepared to explain their algorithms to the public.

"A recurring pattern has developed," says Frank Pasquale at the University of Maryland, "in which some entity complains about a major Internet company's practices, the company claims that its critics don't understand how its algorithms sort and rank content, and befuddled onlookers are left to sift through rival stories in the press." Companies should be prepared to make themselves open to algorithmic audits, as suggested by mathematician and Big Tech critic Cathy O'Neil, in case of complaints or concerns about algorithmic bias that could allow for discrimination in the workplace, healthcare, education, and so on.[7]

Individuals should also have their digital rights legalized. Former *Wired* editor John Battelle has proposed a digital bill of rights that would assign possession of data to its true owner, which is, of course, the user and generator of that data, not the company that made off with it. He believes this notion should be so central that it should be enshrined as an amendment to the Constitution. As the Europeans have put it, people should also have a "right to be forgotten," in which companies must delete any data held on individuals should they wish it. Two million Europeans have already made the choice to opt out. Finally, I would like to see a digital consumer protection bureau, with tough rules around discrimination by algorithms, and a system for ensuring that individuals can access and understand how their personal data is being used, as we can with credit scores today.

All of this relates to the need for more transparency and simplicity in the discussion about Big Tech. Complexity (or the illusion of it) is too often used to avoid legitimate public interest questions, such as how propagandists get their messages across, or how users are tracked and valued. Companies should help us understand by opening the black box of their algorithms. This needn't be a competitive disadvantage; research has shown that it is the amount of data plugged into an algorithm, rather than the cleverness of the algorithm itself, that is the asset. And one could argue that greater transparency is a revenue generator, in that the more users trust what companies are doing, the more willing they may be to part with valuable data.

And the more trusting investors might be of the Big Tech platforms, which have lost so much trust. As one senior policy maker's aide pointed out to me when I began researching this book, data is the most valuable commodity on the planet, and yet companies that traffic in it don't have to declare its value clearly on their financial statements. Currently, the monetary value of data gets shoehorned into "goodwill" on financial statements or, more often, is left out entirely.

That should change, for all kinds of reasons, not the least of which is that investors can't get a remotely accurate picture of what a tech company is worth without understanding the value of exactly what they are trafficking in. (Imagine if you couldn't see the value of the assets held by GM or Ford on their balance sheets.) But even more important is the fact that when we are the product, when our data is what's being collected, we have a right to know how much it's worth—and then decide, as a society, if we should be receiving some of that value ourselves.

We should also consider whether the public sector, rather than private companies, should be the repository of some data wealth, and help ensure that private sector actors have equal access to it and that citizens have more control over just how it is monetized. The conventional wisdom has been that in the brave new world of big data and artificial intelligence, which will drive global growth over the next several decades, there can be only two models: China's surveillance state, in which the government knows and directs all; or the light touch regulation of the United States, which has bred a collection of monopoly powers that may well be choking off job creation and growth in the larger economy.

But there is a third way—one that France and other countries are pursuing—that aims for a middle ground. In Europe, the public sector already holds a large amount of data—in health, transportation, defense, security, and the environment—of the sort that will be needed to develop artificial intelligence and other big data applications. Companies could potentially access these large troves of data held by state institutions, but with public oversight. Citizens would have a say, via elected officials, in the sort of research and big data applications that this data could be used to develop. And companies large and small would have equal access to the gold mine; this would address one of the most frequent complaints I hear from data-driven start-ups in the United States, which is that the biggest players have walled off access to crucial data.

A Fair System of Tax for the Digital Age

Tech, like finance, has hugely benefited from those intangible riches like data and information that can be so easily moved to the tax haven du jour, for the very reason that they are intangible—these assets are virtual rather than physical (like factories or machinery or brick-and-mortar stores) and so can be located anywhere. But the revelations of the Panama Papers, which uncovered how rich companies and individuals around the world were offshoring vast sums of money,[8] have helped to galvanize public debate around how to create a fairer system of tax for the information age, and the United Kingdom, France, India, and others are now suggesting fundamental shifts in the nature of corporate taxation in an effort to level the playing field.

"The current system favors intangible rich firms over those that make money from tangible assets, and multinational firms over small, local companies," says Nobel Prize–winning economist and Columbia University professor Joseph E. Stiglitz, who heads up the Independent Commission for the Reform of International Corporate Taxation, a group of academics and policy makers pushing for global tax reform. "Firms like this can use financial engineering to play all sorts of games," says Stiglitz, who favors a global flat tax on such firms to avoid a zero-sum race to the bottom to the lowest-cost tax havens.[9]

How would this actually work in practice? A key idea found in many of the proposals from tax reform advocates is to tax revenues at the point of sale, rather than profits, which would reduce the sort of financial engineering that allows IP- and data-rich companies like Apple and Google to offshore profits in tax havens like Ireland and the Netherlands.

The problem was first raised at the global level by the Organization for Economic Co-operation and Development in 2012, via its Base Erosion and Profit Shifting initiative; it's now under discussion in forums like the United Nations, the World Bank, and the International Monetary Fund. The issue has been turbocharged by an increasing awareness that the companies that hold the majority of wealth today have no need of a major physical presence in their various markets, or even a fixed national headquarters.

One of the key points that tax reform advocates make is that the labor market disruption caused by Big Tech (which was covered in chapter 8) is forcing states to revamp educational systems, improve vocational training, and invest a lot more to create a twenty-first-century workforce—which, of course, requires tax revenue. But while nearly every country agrees that the current system isn't working, there's not yet consensus on what the new system should be.

The United Kingdom, for example, has said that if there's no international consensus, it plans to unilaterally pass a minimum digital tax. A number of EU finance ministers have come out in support of a tax on revenue versus profits. Other countries, such as India, have already implemented "equalization" levies on payments in excess of $1,500 to foreign enterprises without permanent establishment in the country. This means that when, say, Amazon makes a sale there, a certain amount of tax is withheld on the payment. China and Germany are inclined to buy into the U.S. philosophy of a minimum corporate tax, since they have large companies (which include not just tech firms but automakers) to protect. The United Kingdom and France want to locate value in data and users. And here in the United States, while Trump set a de facto floor on digital tax, he has also inflamed the debate about where value lives in the digital age. All of this points to the fact that in a fractured and politically polarized world, tax on digital goods may become yet another aspect of global trade relations to be weaponized.

Whatever happens, it will represent a big shift in the old order, and the Silicon Valley giants are, of course, complaining bitterly about all of it. At a 2017 OECD conference at the University of California, Berkeley, Robert Johnson, a representative for the Silicon Valley Tax Directors Group, insisted that "raw user data isn't like oil.…Value is created by the development and production of goods and services, not consumption."[10]

Yet, data is *exactly* like oil. In fact, it's even more valuable. Crafting a smart and fair system of digital taxation will not be easy, as the outcry over the various international plans suggests. But at a time when corporations hold more economic power relative to government than ever before, finding a way to reclaim some of that wealth for citizens will be essential to ensuring a functioning democracy.

A Digital New Deal

The prospect of massive technology-related job displacement is a major source of public anxiety about Big Tech—so much so that a relatively unknown entrepreneur named Andrew Yang, the founder of a nonprofit organization that links college graduates to start-up employment, launched a 2020 White House bid on an anti-AI platform. He will not be successful, but the issue—the human cost of artificial intelligence, big data, and automation—will be a major topic in the 2020 U.S. elections. The answer to the question of whether AI will help or hurt workers depends first on your time frame. Technology is always a net job creator over the long run, but, as Keynes put it, in the long run we are all dead.

Perhaps more salient, then, is the second key factor: your socioeconomic class. In the next five years or so, as digital technologies make their way into every industry, they will benefit those at the top with the skills and education to leverage the productivity advantages they afford, and will therefore be likely to increase the winner-takes-all trend in global labor markets. This has massive consequences. While digitalization has the potential to boost productivity and growth, it may also hold back demand if it compresses labor's share of income and increases inequality. One 2018 McKinsey survey of global executives found that the majority believed they would need to retrain or replace more than a quarter of their workforce by 2023 to digitize their businesses. At a conference in that same year, I heard chief executives from large U.S. multinationals discussing ways in which technology would be able to replace 30 to 40 percent of the jobs in their companies over the next few years—and fretting about the political impact of layoffs on that scale.

I would like to propose a radical solution: Do not lay them off.

I am not asking corporate America to keep workers on as charity. I am suggesting that the public and private sector come together in what could be a kind of digital New Deal. As many jobs as will be replaced by automation, there are other areas—customer service, data analysis, and so on—that desperately need talent. Companies that pledge to retain workers and retrain them for new jobs should be offered tax incentives to do so. The United States should take a page out of the post–financial crisis German playbook, in which large-scale layoffs were avoided as both the public and private sector found ways to continue to use labor even as demand dipped. Companies were given government subsidies to keep workers on, and spent the cash on factory upgrades, technical improvements, and training costs, all of which helped German companies grab market share from U.S. rivals in China when growth returned. Corporations also contributed spare workers to public schemes that benefited the larger economy.

In the United States, Cornell law professor Saule Omarova and colleague Robert Hockett have proposed a new National Investment Authority—a hybrid of the New Deal–era Reconstruction Finance Corporation, a modern sovereign wealth fund, and a private equity firm—that would develop and implement a national strategy to remake the real economy for the digital age.

"The proposal is framed in terms of financing public infrastructure, but it is much broader and more ambitious than simply new roads," says Omarova. "We envision it along the lines of the latter-day New Deal approach to financing transformative, large-scale, publicly beneficial projects that would create sustainable jobs and help the country regain its competitive edge—but without exacerbating inequality and excesses of private power. Even though the scope of our proposal is broader than specific AI-related problems, it specifically targets these types of structural imbalances in the economy. We think of this new proposed entity as a 'publicly owned BlackRock' that will finance and channel technological progress in ways that benefit all of us, and not just the richest few."

There are plenty of such projects in the United States that workers could be deployed on now—helping expand rural broadband, for example. The largest companies might even pledge money and excess labor for such a project, which would ultimately supply them with more customers by creating demand in low-growth areas. In a nod to the number of workers who face being downsized, you could call it the 25 percent solution. A way for companies and government to turn a potential employment disaster into an opportunity by using it to train a twenty-first-century workforce and build the public infrastructure to support it. The alternatives, slower growth and more polarized politics, are not pretty.

How to Ensure Our Digital Health and Wellness

This is a hard one, because the technology we've discussed throughout this book has been so pervasive and mind altering. One of the reasons that it's tough to tackle the challenges wrought by Big Tech is that we spend so much time being distracted by it. Fortunately, there are those who have managed to look up for long enough to form a guerrilla movement aimed at pressuring Big Tech to adjust its business model to reduce the human costs of its products. Activists and legislators alike are taking aim at the slot machine aspects of our iPhone addiction, calling for regulation that would protect children from the most noxious types of predatory behavior and marketing online, and looking at whether all of us—kids and adults alike—should be spending a lot less time on our devices.

The short answer—yes. If governments are empowered to restrict mind-altering drugs, then why not limit mind-altering technology, whose effects, being more profound and more ubiquitous, pose a greater hazard? The Food and Drug Administration was established back in 1906 in response to the outrage raised by Upton Sinclair's *The Jungle,* a searing account of the nauseating health hazards created by an unregulated meatpacking industry. I'm hoping that *Don't Be Evil* might help create a similar environment that would foster the creation of a digital FDA for the brain, as the current FDA is for the body. It would study the effects of all the new technology—not just on our own mental health, but the health of the nation—and offer sensible regulations to ensure that the technology that's now so indispensable is in service to us, not betraying us.

Big Tech is, as I said in the first chapter, big. That's one reason why we haven't seen more of these changes, and sooner. The shifts over the past twenty years are so broad and so deep that they are still being metabolized in the public consciousness. Silicon Valley is the richest industry in history, rich enough to buy its way out of quite a lot of trouble. Its products are bright and shiny and life-changing enough that we are all too often willing to settle for the dark trade-offs. That has been the paradox: The good that it does—the information sharing, the relationship building, the productivity enhancing—has been made possible by the bad: the spying, the selling, and the utter breaches of truth and public trust. Because the positives have been so divine—the ability to summon a fact or a cab in the twinkling of an eye—the diabolical negatives have been overlooked.

But it's time to stop being willfully blind. With Big Tech's wealth and power has come tremendous arrogance. There is a sense that society should be reshaped in its image—that we should all be prepared to move faster, work harder, and disrupt anything and everything. But the truth is that Big Tech answers to us—the people. The United States is in danger of becoming an oligopoly run by the wealthiest and most connected, and all too often we feel powerless to change the rules by which companies operate. We need to shake off that feeling of impotence and understand that we can make the rules of the digital economy and society what *we* want and need them to be. What's more, the stakes are too high for us *not* to do that. The history of technology is the history of transformation. And no transformation is ever complete. Industrialization expanded opportunity even as it led to the exploitation of factory workers, which led to government reforms, which gave rise to a backlash in the form of the dominance of Chicago School, neoliberal economic ideas, as well as political libertarianism, both of which have led in turn to some of the excesses of Big Tech. And so it goes.

Big Tech's size and scale and speed have made it difficult to track and control. But we are beginning to understand exactly what we've given up to get all the bright and shiny new things that we have. No new technology remains unchanged, or keeps its hold over the public forever. The railroads once appeared to be an unstoppable force, until wise public officials put them in service to the broader economy rather than merely the robber barons who founded them. Humans are the makers of the new machines, and despite the dystopian paranoia about artificial intelligence, they are still the masters of them. With that power comes the ability and, indeed, the responsibility to select and then create the future we want from Big Tech— for ourselves, and our children.

As for me, it's a future that will include less screen time, and—at least in the short term—a bit more downtime. I've decided that my own mental health depends on checking email less, cutting off most of my social media, and turning my devices off after dinner. The same goes for my kids. I began writing this book after discovering my son had become an online game addict. Since then, we've worked hard to change how he relates to digital media. There's no screen time on school nights, and he's allowed only two-hour windows on weekend days (which means he spends a lot more time reading, as well as playing basketball at the Y). When he is online, I try to be there with him when I can, as a mother, to see what he's doing—and as a journalist, to see what the attention merchants have cooked up next.

Alex still loves his YouTube and his online games.

But not as much as I love parental controls.

Acknowledgments

Many thanks to my editors and colleagues at the *Financial Times* for encouraging me to write and report on the topic of Big Tech in my capacity as global business columnist.

Thanks also to the dozens of sources who contributed their thoughts, experiences, and research to this book. Among the most helpful and supportive were Barry Lynn, Rafi Martina, Frank Pasquale, Jonathan Taplin, Tristan Harris, Roger McNamee, Kiril Sokoloff, Nick Johnson, Rob Johnson, John Battelle, Tim O'Reilly, Shoshana Zuboff, Elvir Causevic, Luther Lowe, Shivaun Raff, Lina Khan, Bill Janeway, B. J. Fogg, Glen Weyl, Luigi Zingales, Michael Wessel, Anya Schiffrin, Joseph E. Stiglitz, David Kappos, James Manyika, George Soros, and David Kirkpatrick.

I also benefited tremendously from reading the work of former colleagues Steven Levy and Brad Stone, as well as other experts, including Jaron Lanier, Frank Foer, Cathy O'Neil, Eric Posner, Hal Varian, Carl Shapiro, Jonathan Haskel, Stian Westlake, Tim Wu, Saule Omarova, Robert C. Hockett, Andrew McAfee, Erik Brynjolfsson, Arun Sundararajan, Viktor Mayer-Schönberger, Kenneth Cukier, Thomas Ramge, Niall Ferguson, and Ken Auletta. At Google, I'd like to particularly thank Corey duBrowa, who endeavored to make it easier for me to ask tough questions, Kent Walker for taking his time to share thoughts on the record, and Karan Bhatia for his insights.

And now, for the personal—a huge thank-you to my husband,

John Sedgwick, and my children, Darya and (of course!) Alex, for bearing with me on yet another book project. Also, thank you to my amazing agent, Tina Bennett, a total rock star who is always three steps ahead of the competition, and my incredibly talented, even-keeled, and hardworking editor Talia Krohn, who not only made my prose ten times better, but gave up many of her own nights and weekends to bring this book in on deadline. Ditto fact checker Julie Tate and research assistant Hannah Assadi, both of whom are top-shelf. And, finally, a huge thanks to Currency publisher Tina Constable and the entire Currency team for believing in me (again) and in the importance of this book. You all are the best in the business, and I feel so grateful to have you in my corner.

Notes

Author's Note

1. McKinsey Global Institute calculations, Rana Foroohar, "Superstar Companies Also Feel the Threat of Disruption," *Financial Times,* October 21, 2018.
2. Jeff Desjardins, "How Google Retains More than 90% of Market Share," *Business Insider,* April 23, 2018.
3. "Facebook by the Numbers: Stats, Demographics, and Fun Facts," Omnicore, January 6, 2019.
4. Celie O'Neil-Hart, "The Latest Video Trends: Where Your Audience Is Watching," Think with Google, April 2016.
5. Sarah Sluis, "Digital Ad Market Soars to $88 Billion, Facebook and Google Contribute 90% of Growth," AdExchanger, May 10, 2018; James Vincent, "99.6 Percent of New Smartphones Run Android or iOS," *Verge,* February 16, 2017.
6. Mark Jamison, "When Did Making Customers Happy Become a Reason for Regulation or Breakup?" AEIdeas, June 8, 2018.
7. "The Regulatory Case Against Platform Monopolies," 13D Research, December 4, 2017.
8. Henry Taylor, "If Social Networks Were Countries, Which Would They Be?" WeForum, April 28, 2016.
9. Michael J. Mauboussin et al., "The Incredible Shrinking Universe of Stocks," Credit Suisse, March 22, 2017.
10. Ian Hathaway and Robert E. Litan, "Declining Business Dynamism in the United States: A Look at States and Metros," Brookings Institution, May 5, 2014.
11. Zoltan Pozsar, "Gobal Money Notes #11," Credit Suisse, January 29, 2018.
12. Mancur Olson, *The Rise and Decline of Nations* (New Haven, Connecticut: Yale University Press, 1982).
13. Rana Foroohar, "Why You Can Thank the Government for Your iPhone," *Time,* October 27, 2015.

14. Author interview with John Battelle in 2017.

15. Rana Foroohar, "Echoes of Wall Street in Silicon Valley's Grip on Money and Power," *Financial Times,* July 3, 2017.

16. Tom Hamburger and Matea Gold, "Google, Once Disdainful of Lobbying, Now a Master of Washington," *The Washington Post,* April 12, 2014.

17. Rana Foroohar, "Silicon Valley Has Too Much Power," *Financial Times,* May 14, 2017; Foroohar, "Echoes of Wall Street in Silicon Valley's Grip."

18. Shoshana Zuboff, *The Age of Surveillance Capitalism: The Fight for a Human Future at the New Frontier of Power* (New York: Public Affairs, 2019), introductory page.

19. Shoshana Zuboff, "Big Other: Surveillance Capitalism and the Prospects of an Information Civilization," *Journal of Information Technology,* April 17, 2015.

20. Niall Ferguson, *The Square and the Tower: Networks and Power, from the Freemasons to Facebook* (New York: Penguin, 2018).

Chapter 1: A Summary of the Case

1. Daisuke Wakabayashi, "Eric Schmidt to Leave Alphabet Board, Ending an Era That Defined Google," *The New York Times,* April 30, 2019.

2. Viktor Mayer-Schönberger and Thomas Ramge, *Reinventing Capitalism in the Age of Big Data* (New York: Basic Books, 2018); Viktor Mayer-Schönberger and Kenneth Cukier, *Big Data: A Revolution That Will Transform How We Live, Work, and Think* (Boston: Houghton Mifflin Harcourt, 2013).

3. While I didn't write about this episode myself at the time, there were a variety of articles published detailing various parts of the meeting, including Hannah Clark's piece "The Google Guys In Davos" (*Forbes,* January 26, 2007), as well as one from the *Financial Times*'s Andrew Edgecliffe-Johnson, with whom I now work ("The Exaggerated Reports of the Death of the Newspaper," *Financial Times,* March 30, 2007).

4. Sheila Dang, "Google, Facebook Have Tight Grip on Growing US Online Ad Market," Reuters, June 5, 2019.

5. Keach Hagey, Lukas I. Alpert, and Yaryna Serkez, "In News Industry, a Stark Divide Between Haves and Have-Nots," *The Wall Street Journal,* May 4, 2019.

6. Judge Richard Leon, memorandum opinion in *United States of America v. AT&T Inc.,* U.S. District Court for the District of Columbia, June 12, 2018.

7. Sheera Frenkel et al., "Delay, Deny, and Deflect: How Facebook's Leaders Fought Through Crisis," *The New York Times,* November 14, 2018.

8. Edelman Trust Barometer, 2018, https://www.edelman.com/trust-barometer.

9. Peter Dizikes, "Study: On Twitter, False News Travels Faster than True Stories," MIT News, March 8, 2018.

10. Federica Cocco, "Most US Manufacturing Jobs Lost to Technology, Not Trade," *Financial Times,* December 2, 2016.

11. "Populist Insurrections: Causes, Consequences, and Policy Reactions," G30 Occasional Lecture, YouTube, April 26, 2017.

12. McKinsey Global Institute, "'Superstars': The Dynamics of Firms, Sectors, and Cities Leading the Global Economy," October 2018.

13. Alex Shephard, "Facebook Has a Genocide Problem," *The New Republic,* March 15, 2018.

14. Edelman Trust Barometer, ibid.

15. Rana Foroohar, "The Dangers of Digital Democracy," *Financial Times,* January 28, 2018.

16. George Soros, "Remarks Delivered at the World Economic Forum," January 24, 2019, https://www.georgesoros.com/2019/01/24/remarks-delivered-at-the-world-economic-forum-2/.

17. Rana Foroohar, "Facebook's Data Sharing Shows It Is Not a US Champion," *Financial Times,* June 6, 2018.

18. Kate Conger and Daisuke Wakabayashi, "Google Employees Protest Secret Work on Censored Search Engine for China," *The New York Times,* August 16, 2018.

19. Foroohar, "Facebook's Data Sharing Shows It Is Not a US Champion."

20. Ahmed Al Omran, "Netflix Pulls Episode of Comedy Show in Saudi Arabia," *Financial Times,* January 1, 2019.

21. Issie Lapowsky, "How the LAPD Uses Data to Predict Crime," *Wired,* May 22, 2018, https://www.wired.com/story/los-angeles-police-department-predictive-policing/.

22. Mark Harris, "If You Drive in Los Angeles, the Cops Can Track Your Every Move," *Wired,* November 13, 2018.

23. Richard Waters, Shannon Bond, and Hannah Murphy, "Global Regulators' Net Tightens Around Big Tech," *Financial Times,* June 6, 2019, page 14.

24. Frenkel et al., "Delay, Deny, and Deflect."

25. Jia Lynn Yang and Nina Easton, "Obama and Google (A Love Story)," *Fortune,* October 26, 2009; Robert Epstein, "How Google Could Rig the 2016 Election," *Politico Magazine,* August 19, 2015; Google Analytics Solutions, "Obama for America Uses Google Analytics to Democratize Rapid, Data-Driven Decision Making," accessed May 9, 2019, https://analytics.google blog.com/.

26. Epstein, "How Google Could Rig the 2016 Election."

27. Sean Gallagher, "Amazon Pitched Its Facial Recognition to ICE, Released Emails Show," Ars Technia, October 24, 2018; Andrea Peterson and Jake Laperruque, "Amazon Pushes ICE to Buy Its Face Recognition Surveillance Tech," Daily Beast, October 23, 2018.

28. Rana Foroohar, "Release Big Tech's Grip on Power," *Financial Times,* June 18, 2017.

29. Ibid.

30. Steven Levy, *In the Plex: How Google Thinks, Works, and Shapes Our Lives* (New York: Simon & Schuster, 2011), 363.

31. ALA News, "Libraries Applaud Dismissal of Google Book Search Case," American Library Association, November 14, 2013.

32. Brody Mullins and Jack Nicas, "Paying Professors: Inside Google's Academic Influence Campaign," *The Wall Street Journal,* July 14, 2017, https://www.wsj.com/articles/paying-professors-inside-googles-academic-influence-campaign-1499785286.

33. Ryan Nakashima, "Google Tracks Your Movements, Like It or Not," Associated Press, August 13, 2018; Sean Illing, "Cambridge Analytica, the Shady Data Firm That Might Be a Key Trump-Russia Link, Explained," Vox, April 4, 2018.

34. Matthew Rosenberg, Nicholas Confessore, and Carole Caldwalladr, "How Trump Consultants Exploited the Facebook Data of Millions," *The New York Times,* March 17, 2018.

35. Camila Domonoske, "Google Announces It Will Stop Allowing Ads for Payday Lenders," NPR, May 11, 2016.

36. Rana Foroohar, "Dangers of Digital Democracy," *Financial Times,* January 28, 2018.

37. Rana Foroohar, "Big Tech Must Pay for Access to America's 'Digital Oil,'" *Financial Times,* April 7, 2019.

38. Jennifer Valentino-DeVries et al., "Your Apps Know Where You Were Last Night, and They're Not Keeping It Secret," *The New York Times,* December 10, 2018.

39. Ben Casselman and Conor Dougherty, "As Investors Flip Housing Markets, Home Buyers Are Reeling," *The New York Times,* June 21, 2019.

40. Terje, "AI Could Add $6 Trillion to the Global Economy," Feelingstream, May 29, 2018.

41. Tim Wu, "In the Grip of the New Monopolists," *The Wall Street Journal,* November 13, 2010.

42. David Z. Morris, "Netflix Is Expected to Spend Up to $13 Billion on Original Programming This Year," *Fortune,* July 8, 2018.

43. "Amazon's Cloud Will Connect Volkswagen's Vast Factory Network," *69News,* WFMZ, March 27, 2019.

44. Angela Chen, "Amazon's Alexa Now Handles Patient Health Information," *The Verge*, April 4, 2019.

45. Bloomberg Billionaires Index, accessed May 9, 2019, https://www.bloom berg.com/billionaires.

46. Lance Whitney, "Apple, Google, Others Settle Antipoaching Lawsuit for $415 Million," https://www.cnet.com/news/apple-google-others-settle-anti -poaching-lawsuit-for-415-million/.

47. Leigh Buchanan, "American Entrepreneurship Is Actually Vanishing. Here's Why," *Inc.,* May 2015.

48. The Hamilton Project, "Start-up Rates Are Declining Across All Sectors," accessed May 9, 2019, http://www.hamiltonproject.org/charts/start_up_rates _are_declining_across_all_sectors.

49. Ian Hathaway and Robert E. Litan, "Declining Business Dynamism in the United States: A Look at States and Metros," Brookings Institution, May 5, 2014.

50. Derek Thompson, "America's Monopoly Problem," *The Atlantic,* October 2016.

51. Kara Swisher, "Is This the End of the Age of Apple?" *The New York Times,* January 3, 2019.

52. Lina M. Khan, "Amazon's Antitrust Paradox," *Yale Law Journal* 126, no. 3 (January 2017).

53. Robert Shapiro and Siddhartha Aneja, "Who Owns Americans' Personal In-formation and What Is It Worth?" Future Majority, March 8, 2019.

54. Foroohar, "Big Tech Must Pay for Access to America's 'Digital Oil.'"

55. Colby Smith, "Peak Buybacks?" *Financial Times,* November 7, 2018.

56. Nico Grant and Ian King, "Big Tech's Big Tax Ruse: Industry Splurges on Buybacks Not Jobs," Bloomberg, April 14, 2019.

57. According to data compiled by the government relations firm Mehlman, Castagnetti, 2019.

58. Cade Metz, "Why WhatsApp Only Needs 50 Engineers for Its 900M Users," *Wired,* September 15, 2015.

59. Alistair Gray, "US Retailers Shut Up Shop as Amazon's March Continues," *Financial Times,* March 8, 2019.

60. James Manyika et al., "Jobs Lost, Jobs Gained: What the Future of Work Will Mean for Jobs, Skills, and Wages," McKinsey and Company, November 2017.

61. "Mapping Inequalities Across the On-Demand Economy," Data and Society, accessed May 9, 2019, https://datasociety.net/initiatives/future-of-labor/ mapping-inequalities-across-the-on-demand-economy/.

62. Shoshana Zuboff, "Big Other: Surveillance Capitalism and the Prospects of an Information Civilization," *Journal of Information Technology,* April 17, 2015.

63. Wikipedia, s.v. "The Great Transformation," last modified March 29, 2019, https://en.wikipedia.org/wiki/The_Great_Transformation_(book).

64. Zuboff, "Big Other," 80.

65. Michael Winnick, "Putting a Finger on Our Phone Obsession," June 16, 2016, https://blog.dscout.com/mobile-touches.

66. Nir Eyal with Ryan Hoover, *Hooked: How to Build Habit-Forming Products* (New York: Portfolio/Penguin, 2014), 1.

67. Rana Foroohar, "All I Want for Christmas Is a Digital Detox," *Financial Times,* December 22, 2017.

68. Rana Foroohar, "Vivienne Ming: 'The Professional Class Is About to Be Blindsided by AI," *Financial Times,* July 27, 2018.

69. Sam Levin, "Facebook Told Advertisers It Can Identify Teens Feeling 'Insecure' and 'Worthless,'" *The Guardian,* May 1, 2017.

70. Foroohar, "All I Want for Christmas Is a Digital Detox."

71. Emily Bary, "Apple Never Meant for You to Spend So Much Time on Your Phone, Tim Cook Says," MarketWatch, June 27, 2019.

72. Olivia Solon, "Ex-Facebook President Sean Parker: Site Made to Exploit Human 'Vulnerability,'" *The Guardian,* November 9, 2017.

Chapter 2: The Valley of the Kings

1. Rana Foroohar and Edward Luce, "Privacy as a Competitive Advantage," *Financial Times,* October 16, 2017.

2. Search Engine Market Share, Statcounter, accessed May 9, 2019, http://gs.statcounter.com/search-engine-market-share/all/worldwide/2009.

3. Rana Foroohar, "Facebook Has Put Growth Ahead of Governance for Too Long," *Financial Times,* December 23, 2018.

4. John Battelle, *The Search: How Google and Its Rivals Rewrote the Rules of Business and Transformed Our Culture* (New York: Portfolio, 2005), 54.

5. Roger McNamee, *Zucked: Waking Up to the Facebook Catastrophe* (New York: Penguin, 2019), 144.

6. Ibid.

7. Sheera Frenkel et al., "Delay, Deny, and Deflect: How Facebook's Leaders Fought Through Crisis," *The New York Times,* November 14, 2018.

8. Tasneem Nashrulla, "A Top George Soros Aide Called for an Independent Investigation of Facebook's Lobbying and PR," BuzzFeed News, November 15, 2018.

9. Jeff Bercovici, "Peter Thiel Wants You to Get Angry About Death," *Inc.,* July 7, 2015; Tad Friend, "Silicon Valley's Quest to Live Forever," *The New Yorker,* March 27, 2017.

10. Marco della Cava et al., "Uber's Kalanick Faces Crisis over 'Baller' Culture," *USA Today,* February 24, 2017.

11. Jeff Bezos, "No Thank You, Mr. Pecker," Medium, February 7, 2019.

12. Daisuke Wakabayashi and Katie Benner, "How Google Protected Andy Rubin, the 'Father of Android,'" *The New York Times,* October 25, 2018.

13. Aarian Marshall, "Elon Musk Reveals His Awkward Dislike of Mass Transit," *Wired,* December 14, 2017.

14. Levy, *In the Plex,* 121.

15. Levy, *In the Plex,* 13.

16. Ken Auletta, "The Search Party," *The New Yorker,* January 6, 2008.

17. Rana Foroohar, "Echoes of Wall Street in Silicon Valley's Grip on Money and Power," *Financial Times,* July 3, 2017.

18. Rana Foroohar, "Big Tech Can No Longer Be Allowed to Police Itself," *Financial Times,* August 27, 2017.

Chapter 3: Advertising and Its Discontents

1. Scott Shane, "These Are the Ads Russia Bought on Facebook in 2016," *The New York Times,* November 1, 2017; Cecilia Kang et al., "Russia-Financed Ad Linked Clinton and Satan," *The New York Times,* November 1, 2017.

2. Indictment, *United States of America v. Internet Research Agency,* U.S. District Court for the District of Columbia, accessed May 9, 2019, https://www.justice.gov/file/1035477/download.

3. Author reporting with Guillaume Chaslot.

4. Max Fisher and Amanda Taub, "On YouTube's Digital Playground, a Gate Left Wide Open for Pedophiles," *The New York Times,* June 4, 2019, page A8.

5. Rob Copeland, "YouTube Weighs Major Changes to Kids' Content Amid FTC Probe," *The Wall Street Journal,* June 19, 2019.

6. Tim Wu, "Aspen Ideas Festival: 'Is the First Amendment Obsolete?'" June 2018.

7. Zeynep Tufekci, "Russian Meddling Is a Symptom, Not the Disease," *The New York Times,* October 3, 2018.

8. Tim Wu, *The Attention Merchants: The Epic Scramble to Get Inside Our Heads* (New York: Knopf, 2016).

9. Craig Silverman, "Apps Installed on Millions of Android Phones Tracked User Behavior to Execute a Multimillion-Dollar Ad Fraud Scheme," BuzzFeed News, October 23, 2018.

10. Rana Foroohar, "Big Tech's Unhealthy Obsession with Hyper-Targeted Ads," *Financial Times,* October 28, 2018; Mark Warner to FTC on Google

Digital Ad Fraud, accessed May 9, 2019, https://www.scribd.com/document/391603927/Senator-Warner-Letter-to-FTC-on-Google-Digital-Ad-Fraud.

11. Steven Levy, *In the Plex: How Google Thinks, Works, and Shapes Our Lives* (New York: Simon & Schuster, 2011), 31.

12. John F. Wasik, "Why Elon Musk Named His Electric Car Tesla," *The Seattle Times,* December 31, 2017.

13. Sergey Brin and Lawrence Page, "The Anatomy of a Large-Scale Hypertextual Web Search Engine," Computer Science Department, Stanford University, 1998.

14. Adam Fisher, "'Google Was Not a Normal Place': Brin, Page and Mayer on the Accidental Birth of the Company That Changed Everything," *Vanity Fair,* July 10, 2018.

15. Ibid.

16. Levy, *In the Plex,* 26.

17. Fisher, "'Google Was Not a Normal Place.'"

18. Ken Auletta, "Searching for Trouble," *The New Yorker,* October 12, 2009.

19. Levy, *In the Plex,* 133.

20. Adam Fisher, *Valley of Genius: The Uncensored History of Silicon Valley (as Told by the Hackers, Founders, and Freaks Who Made It Boom)* (New York: Twelve, 2018).

21. William H. Janeway, *Doing Capitalism in the Innovation Economy* (Cambridge: Cambridge University Press, 2018), 313.

22. Levy, *In the Plex,* 45.

23. Big Easy PowerPoint presentation.

24. Fisher, "'Google Was Not a Normal Place.'"

25. Levy, *In the Plex,* 87.

26. Ibid., 88.

27. John Battelle, *The Search: How Google and Its Rivals Rewrote the Rules of Business and Transformed Our Culture* (New York: Portfolio, 2005), 113–14.

28. David Vise, *The Google Story: Inside the Hottest Business, Media, and Technology Success of Our Time* (New York, Bantam Dell, 2005), 84–85.

29. Fisher, "'Google Was Not a Normal Place.'"

30. Battelle, *The Search,* 125.

Chapter 4: Party Like It's 1999

1. Joshua Cooper Ramo, "Jeffrey Preston Bezos, 1999 Person of the Year," *Time,* December 27, 1999.

2. Wikipedia, graphic of dot-com bubble, accessed May 9, 2019, https://en.wikipedia.org/wiki/Dot-com_bubble#/media/File:Nasdaq_Composite_dot-com_bubble.svg.

3. Simon Dumenco, "Touby Prize," *New York,* July 20, 2007.

4. Rana Foroohar, "Europe's Got Net Fever," *Newsweek International,* September 5, 1999.

5. Ibid.

6. "Dotcom Darlings: Where Are They Now?" *The Telegraph,* accessed May 9, 2019, https://www.telegraph.co.uk/finance/8354329/Dotcom-darlings-where-are-they-now.html/.

7. John Casey, "Accidental Millionaires Sell First Tuesday," *The Guardian,* July 21, 2000.

8. Simon Goodley, "Betfair Buy Spells the Final Flutter," *The Daily Telegraph,* December 22, 2001.

9. Richard Fletcher, "Antfactory Is Wound Up by Shareholders," *The Daily Telegraph,* September 30, 2001.

10. Hal R. Varian, "Economic Scene: Comparing Nasdaq and Tulips Unfair to Flowers," *The New York Times,* February 8, 2001.

11. Olson, *Rise and Decline of Nations.*

12. Rana Foroohar, *Makers and Takers: How Wall Street Destroyed Main Street* (New York: Crown Business, 2016), 130.

13. Ibid.

14. Ibid.

15. Wikipedia, s.v. "Dot-com bubble," last modified May 22, 2019, https://en.wikipedia.org/wiki/Dot-com_bubble.

16. Melanie Warner, "The Beauty of Hype: A Cautionary Tale," *Fortune,* March 1, 1999.

17. Rana Foroohar, "Flight of the Dot-Coms," *Newsweek International,* July 15, 2001.

18. Rana Foroohar and Stefan Theil, "The Dot-Com Witch Hunt," *Newsweek International,* September 3, 2001.

19. Nicole Friedman and Zolan Kanno-Youngs, "Hedge Fund Investor Charles Murphy Dies in Apparent Suicide," *The Wall Street Journal,* March 28, 2017.

20. Rana Foroohar, "Money, Money, Money: Silicon Valley Speculation Recalls Dotcom Mania," *Financial Times,* July 17, 2017.

21. Pan Kwan Yuk and Shannon Bond, "Netflix Returns to Market with $2bn Junk Bond Offering," *Financial Times,* October 22, 2018.

22. Rob Copeland and Eliot Brown, "Palantir Has a $20 Billion Valuation and a Bigger Problem: It Keeps Losing Money," *The Wall Street Journal,* November 12, 2018.

23. Foroohar, "Money, Money, Money."

24. Rana Foroohar, "Another Tech Bubble Could Be About to Burst," *Financial Times,* January 27, 2019.

Chapter 5: Darkness Rises

1. Walter Isaacson, *Steve Jobs* (New York: Simon & Schuster, 2011).
2. Dan Levine, "Apple, Google Settle Smartphone Patent Litigation," Reuters, May 16, 2014.
3. Shanthi Rexaline, "10 Years of Android: How the Operating System Reached 86% Market Share," MSN News, September 25, 2018.
4. Betsy Morris and Deepa Seetharaman, "The New Copycats: How Facebook Squashes Competition from Startups," *The Wall Street Journal*, August 9, 2017.
5. Josh Constine, "Facebook Pays Teens to Install VPN That Spies on Them," TechCrunch, January 29, 2019.
6. Levy, Auletta, and Isaacson all outlined the Apple-Android battle in detail.
7. Steven Levy, *In the Plex: How Google Thinks, Works, and Shapes Our Lives* (New York: Simon & Schuster, 2011), 237–38.
8. Ibid., 80–81.
9. Author interview.
10. Shoshana Zuboff, *The Age of Surveillance Capitalism: The Fight for a Human Future at the New Frontier of Power* (New York: Public Affairs, 2019), 101.
11. Ibid., 63.
12. In a January 2019 interview with Google's chief counsel, Kent Walker, who came on board the company in 2006, he said that Google hadn't begun to think about issues like antitrust until 2008.
13. Rana Foroohar, "Big Tech vs. Big Pharma: The Battle Over US Patent Protection," *Financial Times*, October 16, 2017.
14. In a January 2019 interview with me, Google's chief counsel, Kent Walker, reiterated the patent troll narrative and said he felt the changes to the U.S. patent system had "left us in a better place with a stronger, more robust, more resilient patent system."
15. B. Zorina Khan, "Trolls and Other Patent Inventions: Economic History and the Patent Controversy in the Twenty-First Century," https://papers.ssrn.com/sol3/papers.cfm?abstract_id=2344853.
16. Jaron Lanier, *You Are Not a Gadget: A Manifesto* (New York: Random House, 2011), 125.
17. Foroohar, "Big Tech vs. Big Pharma."
18. Jonathan Taplin's *Move Fast and Break Things* outlines this issue in depth.
19. Levy, *In the Plex*, chapter 7, section 3, covers the book-scanning project.
20. Ibid., 273.
21. Ibid., 350.

22. Ibid., 359.

23. Ibid., 362–63.

24. Taplin, *Move Fast and Break Things*, 260.

25. Per interview with Google executive on background in 2017.

26. Author interview with Taplin in 2017.

27. Author interview with Taplin in 2017; see also *Move Fast and Break Things*.

28. Levy, *In the Plex*, 251.

29. Taplin, *Move Fast and Break Things*, 127–28.

30. Wikipedia, s.v. "Directive on Copyright in the Digital Single Market," last modified May 19, 2019, https://en.wikipedia.org/wiki/Directive_on_Copyright_in_the_Digital_Single_Market.

31. Mehreen Khan and Tobias Buck, "European Parliament Backs Overhaul of EU Copyright Rules," *Financial Times,* March 26, 2019.

32. "'Purchased Protest' Bombshell: Germany's FAZ News Uncovers the Seamy Underbelly of Google's Article 13 Lobbying," Music Technology Policy, March 16, 2019.

33. Khan and Buck, "European Parliament Backs Overhaul of EU Copyright Rules."

34. Editorial Board, "EU Copyright Reforms Are Harsh but Necessary," *Financial Times*, March 26, 2019.

35. Pew Research Center, Newspaper Fact Sheet, June 13, 2018.

36. Rana Foroohar, "A Better US Patent System Will Spur Innovation," *Financial Times,* September 3, 2017.

37. Lance Whitney, "Apple, Google, Others Settle Antipoaching Lawsuit for $415 Million," CNET News, September 3, 2015.

38. Dan Levine, "Apple, Google Agree to Settle Lawsuit Alleging Hiring, Salary Conspiracy," *The Washington Post,* April 24, 2014.

39. Author interview with Peter Harter, 2017.

40. James Thomson, "Tech Giants Buy Start-ups to Kill Competition, Kenneth Rogoff Tells Summit," *Financial Review,* March 7, 2018.

41. Olivia Solon, "As Tech Companies Get Richer, Is It 'Game-Over' for Start-ups?" *The Guardian,* October 20, 2017.

42. Marc Doucette, "Visualizing Major Tech Acquisitions," Visual Capitalist, July 24, 2018.

43. "American Tech Giants Are Making Life Tough for Startups," *The Economist,* June 2, 2018.

44. Ken Auletta, *Googled: The End of the World as We Know It* (New York: Penguin Books, 2009), 110.

Chapter 6: A Slot Machine in Your Pocket

1. Wesley Yin-Poole, "FIFA Player Uses GDPR to Find Out Everything EA Has on Him, Realises He's Spent over $10,000 in Two Years on Ultimate Team," Eurogamer, July 25, 2018.

2. Biographical information on Fogg can be accessed at https://www.bjfogg .com/.

3. Rana Foroohar, "Silicon Valley Has Too Much Power," Financial Times, May 14, 2017.

4. Author interview with Fogg, August 14, 2018.

5. "Slot Machine: The Crack Cocaine of Gambling Addiction," KS Problem Gaming, http://www.ksproblemgambling.org/html/slot_machine.html.

6. Author interviews with Harris.

7. Author interview with Fogg, 2018.

8. Ibid.

9. Hannah Kuchler, "How Facebook Grew Too Big to Handle," Financial Times, March 28, 2019.

10. Author interviews with Fogg, 2017, 2018.

11. Wikipedia, s.v. "B. J. Fogg," last modified February 5, 2019, https:// en.wikipedia.org/wiki/B._J._Fogg.

12. Bianca Bosker, "The Binge Breaker," The Atlantic, November 2016.

13. Ibid.

14. Tristan Harris, "How Technology Is Hijacking Your Mind—from a Magician and Google Design Ethicist," Medium, May 18, 2016. Other information about Harris can be accessed at his Time Well Spent website, http:// www.tristanharris.com/tag/time-well-spent.

15. Michael Winnick, "Putting a Finger on Our Phone Obsession," dscout blog, June 16, 2016.

16. Tiffany Hsu, "Video Game Addiction Tries to Move from Basement to Doctor's Office," The New York Times, June 17, 2018.

17. Betsy Morris, "How Fortnite Triggered an Unwinnable War Between Parents and Their Boys," The Wall Street Journal, December 21, 2018.

18. "The Impact of Media Use and Screen Time on Children, Adolescents, and Families," American College of Pediatricians, November 2016.

19. Jean M. Twenge, "Have Smartphones Destroyed a Generation?" The Atlantic, September 2017; Richard Freed, "The Tech Industry's War on Kids," Medium, March 12, 2018.

20. Darren Davidson, "Facebook Targets 'Insecure' to Sell Ads," The Australian, May 1, 2017.

21. "Over a Dozen Children's and Consumer Advocacy Organizations Request

Federal Trade Commission to Investigate Facebook for Deceptive Practices," Common Sense Media, February 21, 2019.

22. Kristen Duke et al., "Having Your Smartphone Nearby Takes a Toll on Your Thinking," *Harvard Business Review,* March 20, 2018.

23. Child health data can be accessed at the Data Resource Center for Child and Adolescent Health, http://childhealthdata.org/learn-about-the-nsch/NSCH/data.

24. Casey Schwartz, "Finding It Hard to Focus? Maybe It's Not Your Fault," *The New York Times,* August 14, 2018.

25. Nellie Bowles, "A Dark Consensus About Screens and Kids Begins to Emerge in Silicon Valley," *The New York Times,* October 26, 2018.

26. Author interview with Harris, 2017.

27. Rana Foroohar, "The Coming Corporate Crackdown," *Time,* June 3, 2013.

28. Wikipedia, s.v. "Marshall McLuhan," last modified May 9, 2019, https://en.wikipedia.org/wiki/Marshall_McLuhan.

29. Bianca Bosker, "The Binge Breaker," *The Atlantic,* November 2016.

30. Author interviews with Harris, 2017, 2018.

31. Kevin Webb, "The FTC Will Investigate Whether a Multibillion-Dollar Business Model Is Getting Kids Hooked on Gambling Through Video Games," *Business Insider,* November 28, 2018.

32. Tim Bradshaw and Hannah Kuchler, "Smartphone Addiction: Big Tech's Balancing Act on Responsibility over Revenue," *Financial Times,* July 23, 2018.

33. Valentino-DeVries et al., "Your Apps Know Where You Were Last Night, and They're Not Keeping It Secret."

34. David Benoit, "iPhones and Children Are a Toxic Pair, Say Two Big Apple Investors," *The Wall Street Journal,* January 7, 2018.

35. Apple, "iOS 12 Introduces New Features to Reduce Interruptions and Manage Screen Time," June 4, 2018, https://www.apple.com/newsroom/2018/06/ios-12-introduces-new-features-to-reduce-interruptions-and-manage-screen-time/.

36. Rana Foroohar, "Big Tech's Unhealthy Obsession with Hyper-Targeted Ads," *Financial Times,* October 28, 2018.

37. Author interview with Chaslot in 2018.

Chapter 7: The Network Effect

1. Adam Satariano and Mike Isaac, "Facebook Used People's Data to Favor Certain Partners and Punish Rivals, Documents Show," *The New York Times,* December 5, 2018.

2. Ibid.

3. Jonathan Haskel and Stian Westlake, *Capitalism Without Capital: The Rise of the Intangible Economy* (Princeton, N.J.: Princeton University Press, 2018).

4. Conor Dougherty, "Inside Yelp's Six-Year Grudge Against Google," *The New York Times,* January 7, 2017.

5. Author interviews conducted with Lowe, 2017–19.

6. Author interviews with Lowe in 2017–18; Charles Arthur, "Why Google's Struggles with the EC—and FTC—Matter," *Overspill,* April 7, 2015.

7. Brody Mullins, Rolfe Winkler, and Brent Kendall, "Inside the Antitrust Probe of Google," *The Wall Street Journal,* March 19, 2015.

8. Leaked FTC document, page 20. Document can be accessed here: http://graphics.wsj.com/google-ftc-report/img/ftc-ocr-watermark.pdf.

9. Ibid., fn. 12.

10. Leaked FTC document, 26.

11. Ibid.

12. Rana Foroohar, "Google Versus Orrin Hatch," *Financial Times,* September 3, 2018. Varian quote from *The Wall Street Journal,* Ibid.

13. Nitasha Tiku, "How Google Influences the Conversation in Washington," *Wired,* March 13, 2019.

14. Author interview with Walker in January 2019.

15. Madeline Jacobson, "How Far Down the Search Engine Results Page Will Most People Go?" Leverage Marketing, 2015.

16. For general information about antitrust lawsuits, see Wikipedia, s.v. "United States v. Terminal R.R. Ass'n," last modified May 7, 2019, https://en.wikipedia.org/wiki/United_States_v._Terminal_R.R._Ass%27n.

17. "United States v. Reading Co.," https://casetext.com/case/united-states-v-reading-co.

18. Charles Francis Adams Jr., *Railroads: Their Origins and Problems* (1878).

19. Rana Foroohar, "Big Tech Is America's New 'Railroad Problem,' " *Financial Times,* June 16, 2019.

20. Author interview with Walker in 2019.

21. Charles Duhigg, "The Case Against Google," *The New York Times,* February 20, 2018.

22. Ibid.

23. Open Letter to Commissioner Vestager from 14 European CSSs, November 22, 2018, http://www.searchneutrality.org/google/comparison-shopping-services-open-letter-to-commissioner-vestager.

24. William A. Galston and Clara Hendrickson, "A Policy at Peace with Itself: Antitrust Remedies for Our Concentrated, Uncompetitive Economy," Brookings Institution, January 5, 2018.

25. Rana Foroohar, "The Rise of the Superstar Company," *Financial Times,* January 14, 2018.
26. Author interview with James Manyika of the McKinsey Global Institute.
27. Jason Furman, "Productivity, Inequality, and Economic Rents," *The Regulatory Review,* June 13, 2016.
28. David Autor et al., "The Fall of the Labor Share and the Rise of Superstar Firms," NBER Working Paper 23396, National Bureau of Economic Research, May 1, 2017.
29. McKinsey Global Institute, "A New Look at the Declining Labor Share of Income in the United States," May 2019.
30. Foroohar, "The Rise of the Superstar Company."
31. Dan Andrews et al., "Going Digital: What Determines Technology Diffusion Among Firms?" OECD background paper, Third Annual Conference of the Global Forum on Productivity, Ottawa, Canada, June 28–29, 2018.
32. James Manyika et al., "'Superstars': The Dynamics of Firms, Sectors, and Cities Leading the Global Economy," McKinsey Global Institute, October 2018.
33. Haskel and Westlake, *Capitalism Without Capital.*
34. Rana Foroohar, "Superstar Companies Also Feel the Threat of Disruption," *Financial Times,* October 21, 2018.
35. "Autonomous Cars: Self-Driving the New Auto Industry Paradigm," Morgan Stanley Blue Paper, November 6, 2013.
36. Nicholas L. Johnson and Alex Moazed, *Modern Monopolies: What It Takes to Dominate the 21st Century Economy* (New York: St. Martin's Press, 2016).
37. Andrew Hill, "Inside Nokia: Rebuilt from Within," *Financial Times,* April 13, 2011.
38. Steven Levy, *In the Plex: How Google Thinks, Works, and Shapes Our Lives* (New York: Simon & Schuster, 2011), 117.
39. Ibid., 12.
40. Ibid., 14.
41. Ibid., 202.
42. Shoshana Zuboff, "Big Other: Surveillance Capitalism and the Prospects of an Information Civilization," *Journal of Information Technology,* April 17, 2015.
43. Ibid., 15.
44. Rana Foroohar, "The End of Privacy," *Financial Times,* October 29, 2018.
45. Rana Foroohar, "Privacy Is a Competitive Advantage," *Financial Times,* October 15, 2017.

Chapter 8: The Uberization of Everything

1. Leslie Hook, "Uber: The Crisis Inside the 'Cult of Travis,'" *Financial Times,* March 9, 2017.

2. Video of Kalanick arguing with an Uber driver over fares can be accessed here: https://www.youtube.com/watch?v=gTEDYCkNqns.

3. Katy Steinmetz and Matt Vella, "Uber Fail: Upheaval at the World's Most Valuable Startup Is a Wake-Up Call for Silicon Valley," *Time*, June 15, 2017.

4. Sheelah Kolhatkar, "At Uber, a New CEO Shifts Gears," *The New Yorker*, March 30, 2018.

5. Hook, "Uber."

6. Eric Newcomer, Sonali Basak, and Sridhar Natarajan, "Uber's Blame Game Focuses on Morgan Stanley After Shares Drop," *Bloomberg Businessweek*, May 20, 2019.

7. Rana Foroohar, "Travis Kalanick: With His $62.5 Billion Startup, the Uber Founder Is Changing the Nature of Work," *Time*, 2015.

8. Theron Mohamed, "Uber Is Paying Drivers up to $40,000 Each to Celebrate Its IPO," *Markets Insider*, April 26, 2019.

9. Alex Rosenblat, *Uberland: How Algorithms Are Rewriting the Rules of Work* (Oakland: University of California Press, 2018), 5.

10. Ibid., 98, 203.

11. Rob Wile, "Here's How Much Lyft Drivers Really Make," CNN Money, July 11, 2017.

12. Josh Zumbrun, "How Estimates of the Gig Economy Went Wrong," *The Wall Street Journal*, January 7, 2019.

13. Aimee Picchi, "Inside an Amazon Warehouse: Treating Human Beings as Robots," CBS News, April 19, 2018.

14. Michael Sainato, "Accidents at Amazon: Workers Left to Suffer After Warehouse Injuries," *The Guardian*, July 30, 2018.

15. Foroohar, "Vivienne Ming."

16. Jodi Kantor, "Working Anything but 9 to 5," *The New York Times*, August 13, 2014.

17. Rosenblat, *Uberland*, 177.

18. Ibid., 110.

19. "Prediction: How AI Will Affect Business, Work, and Life," *Managing the Future of Work*, Harvard Business School podcast, May 8, 2019, https://www.hbs.edu/managing-the-future-of-work/podcast/Pages/default.aspx.

20. Claudia Goldin and Lawrence F. Katz, *The Race Between Education and Technology* (Cambridge, Mass.: Belknap Press of Harvard University Press, 2008).

21. World Trade Organization, "Impact of Technology on Labour Market Outcomes," 2017.

22. Rana Foroohar, "Gap Between Gig Economy's Winners and Losers Fuels Populists," *Financial Times*, May 2, 2017.

23. International Monetary Fund, "World Economic Outlook, April 2017: Gaining Momentum?" April 2017.

24. Rana Foroohar, "Silicon Valley 'Superstars' Risk a Populist Backlash," *Financial Times,* April 23, 2017.

25. Rana Foroohar and Edward Luce, "The Tech Effect," *Financial Times,* January 15, 2018.

26. Rana Foroohar, "U.S. Capital Expenditure Boom Fails to Live Up to Promises," *Financial Times,* November 25, 2018.

27. Rana Foroohar, "Vivienne Ming"; Pablo Illanes et al., "Retraining and Reskilling Workers in the Age of Automation," McKinsey Global Institute, January 2018.

28. Rana Foroohar, "The 'Haves and Have-Mores' in Digital America," *Financial Times,* August 6, 2017.

29. Foroohar, "Gap Between Gig Economy's Winners and Losers Fuels Populists."

30. Rana Foroohar, "The Rise of the Superstar Company," *Financial Times,* January 14, 2018.

31. Gillian Tett, "Tech Lessons from Amazon's Battle in Seattle," *Financial Times,* May 17, 2018.

32. Christina Warren, "A Brief History of Uber and Google's Very Complicated Relationship," Gizmodo, February 24, 2017.

33. Brian M. Rosenthal, "Taxi Drivers Fell Prey While Top Officials Counted the Money," *The New York Times,* May 20, 2019.

34. Author interview with Schmidt, 2015.

35. John Gapper, "Car Ownership May Peak but Traffic Is on the Rise," *Financial Times,* October 24, 2018.

36. Rana Foroohar, "Strong Unions Will Boost America's Economy," *Financial Times,* July 31, 2017.

37. Foroohar, "Travis Kalanick."

Chapter 9: The New Monopolists

1. Barry C. Lynn, *End of the Line: The Rise and Coming Fall of the Global Corporation* (New York: Doubleday, 2005).

2. Kenneth P. Vogel, "Google Critic Ousted from Think Tank Funded by Tech Giant," *The New York Times,* August 30, 2017.

3. Rana Foroohar, "Lina Khan: 'This Isn't Just About Antitrust. It's About Values,'" *Financial Times,* March 29, 2019.

4. Lina M. Khan, "Amazon's Antitrust Paradox," *Yale Law Journal* 126, no. 3 (January 2017).

5. A good roundup of some of the most compelling research can be found in Jonathan Tepper, with Denise Hern, *The Myth of Capitalism: Monopolies and the Death of Competition* (Hoboken, N.J.: John Wiley and Sons, 2019).

6. Foroohar, "Lina Khan."

7. Brad Stone, *The Everything Store: Jeff Bezos and the Age of Amazon* (New York: Little, Brown, 2013).

8. David Streitfeld, "A New Book Portrays Amazon as Bully," *The New York Times,* October 22, 2013.

9. Sam Moore, "Amazon Commands Nearly Half of Consumers' First Product Search," Bloomreach, October 6, 2015.

10. Khan, "Amazon's Antitrust Paradox."

11. John Koetsier, "Research Shows Amazon Echo Owners Buy 29% More from Amazon," *Forbes,* May 30, 2018.

12. Shapiro and Aneja, "Who Owns Americans' Personal Information and What Is It Worth?"

13. Rana Foroohar, "How Much Is Your Data Worth?" *Financial Times,* April 8, 2019.

14. "Secret of Googlenomics: Data-Fueled Recipe Brews Profitability," *Wired,* May 22, 2009.

15. Paul W. Dobson, "The Waterbed Effect: Where Buying and Selling Power Come Together," *Wisconsin Law Review,* January 2018.

16. Angus Loten and Adam Janofsky, "Sellers Need Amazon But at What Cost?" *The Wall Street Journal,* January 14, 2015.

17. Khan, "Amazon's Antitrust Paradox."

18. Barry C. Lynn and Lina Khan, "The Slow-Motion Collapse of American Entrepreneurship," *Washington Monthly,* July/August 2012.

19. "The Next Capitalist Revolution," *The Economist,* November 15, 2018.

20. David Carr, "How Good (or Not Evil) Is Google?" *The New York Times,* June 21, 2009.

21. Adam Candeub, "Behavioral Economics, Internet Search, and Antitrust," *ISJLP* 9, no. 407 (2014), https://digitalcommons.law.msu.edu/cgi/viewcontent.cgi?article=1506&context=facpubs.

22. David Leonhardt, "The Monopolization of America," *The New York Times,* November 25, 2018.

23. Wu, *Curse of Bigness,* 45.

24. Ibid.

25. Author conversation with Khan in 2018.

26. Foroohar, "Lina Khan."

27. "Next Capitalist Revolution," *The Economist.*

28. Foroohar, "Lina Khan."

29. Rana Foroohar, "Antitrust Policy Is Ripe for a Rethink," *Financial Times,* January 24, 2018.

30. Todd Spangler, "Cord Cutting Explodes: 22 Million U.S. Adults Will Have Canceled Cable, Satellite TV by End of 2017," *Variety,* September 13, 2017.

31. For information concerning the European Union case against Google, see Wikipedia, s.v. "European Union v. Google," last modified May 31, 2019, https://en.wikipedia.org/wiki/European_Union_vs._Google; "Antitrust: Commission Fines Google €4.34 Billion for Illegal Practices Regarding Android Mobile Devices to Strengthen Dominance of Google's Search Engine," European Commission Press Release, July 18, 2018.

32. Author interview with Delrahim in 2018.

33. Author interview, 2018.

34. McNamee, *Zucked,* 285–86.

35. Wu, *After Consumer Welfare, Now What? The "Protection of Competition" Standard Practice,* Competition Policy International, 2018, Columbia Public Law Research Paper, no. 14–608 (2018).

Chapter 10: Too Fast to Fail

1. Robert Lenzner and Stephen S. Johnson, "Seeing Things as They Really Are," *Forbes,* March 10, 1997.

2. For information concerning stock buybacks, see "$407 Billion in Corporate Stock Buybacks! How Are Businesses in Your State Spending the Trump Tax Cuts?" Americans for Tax Fairness press release, May 10, 2018, https://americansfortaxfairness.org/wp-content/uploads/20180510-TTCT-Updates-Release.pdf.

3. Ibid.

4. "Risks Rising in Corporate Debt Market," OECD Report, February 25, 2019.

5. Rana Foroohar, "Apple Sows Seeds of Next Market Swing," *Financial Times,* May 13, 2018.

6. Martin Wolf, "Taming the Masters of the Tech Universe," *Financial Times,* November 14, 2017.

7. Rana Foroohar, "Tech Companies Are the New Investment Banks," *Financial Times,* February 11, 2018.

8. Edelman Trust Barometer 2018, 2019.

9. Rana Foroohar, "Political Ads on Facebook Recall Memories of the Banking Crisis," *Financial Times,* October 2, 2017.

10. Gabriel J. X. Dance et al., "As Facebook Raised a Privacy Wall, It Carved an Opening for Tech Giants," *The New York Times,* December 18, 2018.

11. Rana Foroohar, *Makers and Takers: How Wall Street Destroyed Main Street* (New York: Crown Business, 2016).

12. Martin Wolf, "We Must Rethink the Purpose of the Corporation," *Financial Times,* December 11, 2018.

13. Foroohar, *Makers and Takers.*

14. Douglas Edwards, *I'm Feeling Lucky: The Confessions of Google Employee Number 59* (Boston: Houghton Mifflin Harcourt, 2011), 291.

15. Foroohar, "How Much Is Your Data Worth?"

16. Rana Foroohar, "Facebook Has Put Growth Ahead of Governance for Too Long," *Financial Times,* December 23, 2018.

17. Eric Schmidt and Jared Cohen, *The New Digital Age: Transforming Nations, Businesses, and Our Lives* (New York: Vintage Books, 2014), 261.

18. Rana Foroohar, "It Is Time for a Truly Free Market," *Financial Times,* March 31, 2019.

19. Rana Foroohar, "U.S. Capital Expenditure Boom Fails to Live Up to Promises," *Financial Times,* November 25, 2018.

20. Foroohar, "Tech Companies Are the New Investment Banks."

21. Rana Foroohar, "Banks Jump on the Fintech Bandwagon," *Financial Times,* September 16, 2018; Mark Bergen and Jennifer Surane, "Google and Mastercard Cut a Secret Ad Deal to Track Retail Sales," Bloomberg, August 30, 2018.

22. Stacy Mitchell and Olivia LaVecchia, "Report: Amazon's Next Frontier: Your City's Purchasing," Institute for Self-Reliance, July 10, 2018.

23. Lina M. Khan, "A Separation of Platforms and Commerce," Columbia Law Review, https://columbialawreview.org/content/the-separation-of-platforms-and-commerce/.

24. Foroohar, *Makers and Takers,* 189.

25. Saule T. Omarova, "New Tech v. New Deal: Fintech as a Systemic Phenomcnon," *Yale Journal on Regulation* 36, no. 2 (August 1, 2018).

26. "IMF Warns of Giant Tech Firms' Dominance," BBC News, June 8, 2019.

27. Cathy O'Neil, *Weapons of Math Destruction: How Big Data Increases Inequality and Threatens Democracy* (New York: Crown, 2016), 143–44.

28. Agustín Carstens, "Big Tech in Finance and New Challenges for Public Policy," keynote address at the FT Banking Summit, London, December 4, 2018.

29. Rana Foroohar, "Political Ads on Facebook Recall Memories of the Banking Crisis."

30. Wolf, "Taming the Masters of the Tech Universe."

Chapter 11: In the Swamp

1. Frank Pasquale, *The Black Box Society: The Secret Algorithms That Control Money and Information* (Cambridge, Mass.: Harvard University Press, 2015), 196.
2. Hamburger and Gold, "Google, Once Disdainful of Lobbying, Now a Master of Washington."
3. Pinar Akman, "The Theory of Abuse in Google Search: A Positive and Normative Assessment Under EU Competition Law," *Journal of Law, Technology and Policy* 2017, no. 2 (July 19, 2016): 301–74.
4. "Google Academics Inc.," Google Transparency Project, July 22, 2017, accessed May 9, 2019, https://googletransparencyproject.org/articles/google-academics-inc.
5. Brody Mullins and Jack Nicas, "Paying Professors: Inside Google's Academic Influence Campaign," *The Wall Street Journal,* July 14, 2017.
6. "Google's Silicon Tower," Campaign for Accountability Report, July 19, 2016.
7. Author interview with the aide, 2017.
8. "Does America Have a Monopoly Problem?" U.S. Senate Judiciary Committee hearing, Subcommittee on Antitrust, Competition Policy, and Consumer Rights, March 5, 2019.
9. Nitasha Tiku, "How Google Influences the Conversation in Washington," *Wired,* March 13, 2019.
10. Numbers provided by the Center for Responsive Politics.
11. Tiku, "How Google Influences the Conversation in Washington."
12. David McCabe and Erica Pandey, "Explore Amazon's Wide Washington Reach," Axios, March 13, 2019.
13. Beejoli Shah and Christopher Stern, "How Netflix Scaled Back U.S. Lobbying to Focus on Europe," *The Information,* May 7, 2019.
14. Nicholas Thompson and Fred Vogelstein, "15 Months of Fresh Hell Inside Facebook," *Wired,* May 2019.
15. Philipp Schindler, "The Google News Initiative: Building a Stronger Future for News," March 20, 2018, https://blog.google/outreach-initiatives/google-news-initiative/announcing-google-news-initiative/.
16. Rana Foroohar, "Travis Kalanick: With His $62.5 Billion Startup, the Uber Founder Is Changing the Nature of Work," *Time,* 2015.
17. Author interview with senior aide to Democratic senator.
18. Daniel Kreiss and Shannon C. McGregor, "Technology Firms Shape Political Communication: The Work of Microsoft, Facebook, Twitter, and Google with Campaigns During the 2016 U.S. Presidential Cycle," *Journal of Political Communication* 35, no. 2 (2018).

19. Matt Warman, "Google, Caffeine, and the Future of Speech," *Telegraph*, June 10, 2010.

20. Steven Levy, *In the Plex: How Google Thinks, Works, and Shapes Our Lives* (New York: Simon & Schuster, 2011), 333.

21. Ibid., 334.

22. "Mission Creep-y," Public Citizen, November 2014.

23. Federal Trade Commission, "FTC Charges Deceptive Privacy Practices in Google's Rollout of Its Buzz Social Network," press release, March 30, 2011.

24. "Updating Our Privacy Policies and Terms of Service," Google: Official Blog, January 24, 2012, https://googleblog.blogspot.com/2012/01/updating-our -privacy-policies-and-terms.html.

25. Motion for Temporary Restraining Order and Preliminary Injunction, *Electronic Privacy Information Center v. The Federal Trade Commission*, U.S. District Court for the District of Columbia, February 8, 2012, https://epic .org/privacy/ftc/google/TRO-Motion-final.pdf.

26. "Estimated Total Conversions: New Insights for the Multi-Screen World," *Google Inside Adwords*, October 1, 2013, https://adwords.googleblog .com/2013/10/estimated-total-conversions.html.

27. Federal Trade Commission, "Google Will Pay $22.5 Million to Settle FTC Charges It Misrepresented Privacy Assurances to Users of Apple's Safari Internet Browser," press release, August 9, 2012.

28. Tiku, "How Google Influences the Conversation in Washington."

29. Yang and Easton, "Obama and Google (A Love Story)."

30. Rana Foroohar, "Why Big Tech Wants to Keep the Net Neutral," *Financial Times*, December 17, 2017.

31. Rana Foroohar, "Back to My Roots," *Financial Times*, September 17, 2018.

32. Cecilia Kang, "Net Neutrality Vote Passes House, Fulfilling Promise by Democrats," *The New York Times*, April 10, 2018.

33. Kiran Stacey, "Broadband Groups Cut Capital Expenditure Despite Net Neutrality Win," *Financial Times*, February 7, 2019.

34. "Don't Forget the 'Net Neutrality' Panic," *The Wall Street Journal*, editoral page, June 15–16, 2019.

35. Consumer Watchdog, "How Google's Backing of Backpage Protects Child Sex Trafficking," report from Consumer Watchdog, Faith and Freedom Coalition, Trafficking America Taskforce, DeliverFund, and the Rebecca Project, May 17, 2017.

36. Nicholas Kristof, "Google and Sex Traffickers Like Backpage.com," *The New York Times*, September 7, 2017.

37. Consumer Watchdog, "How Google's Backing of Backpage Protects Child Sex Trafficking."

38. Kieren McCarthy, "Google Lobbies Hard to Derail New US Privacy Laws—Using Dodgy Stats," The Register, March 26, 2018.

39. "Platform Monopolies in NAFTA—The Body Camera Monopoly—Price Discrimination in the Airline Industry," Open Market Institute, May 17, 2018.

40. Rana Foroohar, "Fear and Loathing in Silicon Valley," *Financial Times,* July 23, 2018.

41. Author interview with David Greene.

42. Author interviews with diplomats in Brussels and Washington.

43. Germán Gutiérrez and Thomas Philippon, "How EU Markets Became More Competitive Than U.S. Markets: A Study of Institutional Drift," NBER Working Paper 24700, June 2018, National Bureau of Economic Research.

44. John Paul Rathbone, "Google Strikes Deal to Bring Faster Web Content to Cuba," *Financial Times,* March 28, 2019.

45. Rana Foroohar, "It Is Time for a Truly Free Market," *Financial Times*, March 31, 2019.

Chapter 12: 2016: The Year It All Changed

1. Sean J. Miller, "Digital Ad Spending Tops Estimates," Campaign and Elections, January 4, 2017.

2. Status Memo from Teddy Goff to Clinton Campaign Officials can be accessed here: https://wikileaks.org/podesta-emails/fileid/12403/3324.

3. Kreiss and McGregor, "Technology Firms Shape Political Communication."

4. Ibid., 415.

5. Evan Osnos, "Can Mark Zuckerberg Fix Facebook Before It Breaks Democracy?" *The New Yorker,* September 17, 2018.

6. Joshua Green and Sasha Issenberg, "Inside the Trump Bunker, With 12 Days to Go," Bloomberg, October 27, 2016.

7. Mueller, Robert S., III, "Report on the Investigation into Russian Interference in the 2016 Presidential Election," Homeland Security Digital Library, March 2019, https://www.hsdl.org/?abstract&did=824221.

8. Osnos, "Can Mark Zuckerberg Fix Facebook Before It Breaks Democracy?"

9. McNamee, *Zucked,* 7–8.

10. "Disinformation and 'Fake News': Final Report," United Kingdom Parliament, Digital, Culture, Media and Sport Committee, February 18, 2019.

11. Roger McNamee, "Ever Get the Feeling You're Being Watched?" *Financial Times,* February 7, 2019.

12. Rana Foroohar, "Have You Been Zucked?" *Financial Times,* February 4, 2019.

13. McNamee, "Ever Get the Feeling You're Being Watched?"

14. Amarendra Bhushan Dhiraj, "Report: Facebook's Annual Revenue from 2009 to 2018," CEO World, February 4, 2019.

15. Edward Luce and Rana Foroohar, "Election Manipulation Edition," *Financial Times,* February 19, 2018.

16. Indictment, *United States of America v. Internet Research Agency.*

17. Ryan Mac et al., "Growth at Any Cost: Top Facebook Executive Defended Data Collection in 2016 Memo—and Warned That Facebook Could Get People Killed," BuzzFeed News, March 29, 2018.

18. Osnos, "Can Mark Zuckerberg Fix Facebook Before It Breaks Democracy?"

19. Eli Pariser, "Beware Online 'Filter Bubbles,'" TED Talk, March 2011.

20. McNamee, *Zucked,* 152.

21. Sam Levin, "ACLU Finds Social Media Sites Gave Data to Company Tracking Black Protesters," *The Guardian,* October 11, 2016.

22. Shapiro and Aneja, "Who Owns Americans' Personal Information and What Is It Worth?"

23. Ibid.

24. Ibid.

25. Rana Foroohar, "Companies Are the Cops in Our Modern-Day Dystopia," *Financial Times,* May 27, 2018.

26. Sarah Brayne, "Big Data Surveillance: The Case of Policing," *American Sociological Review* 82, no. 5 (2017).

27. Aria Bendix, "Activists Say Alphabet's Planned Neighborhood in Toronto Shows All the Warning Signs of Amazon HQ2-Style Breakup," *Business Insider,* April 14, 2019.

28. Marco Chown Oved, "Google's Sidewalk Labs Plans Massive Expansion to Waterfront Vision," *Toronto Star,* February 14, 2019.

29. Anna Nicolaou, "Future Shock: Inside Google's Smart City," *Financial Times,* March 22, 2019.

30. Ryan Gallagher, "Google Dragonfly," *Intercept,* March 27, 2019.

31. Shannon Vavra, "Declassified Cable Estimates 10,000 Killed at Tiananmen Square," Axios, December 24, 2017.

32. Matt Sheehan, "How Google Took On China—and Lost," *MIT Technology Review,* December 18, 2018.

33. Mark Warner, "Warner, Colleagues Raise Concerns About Google's Reported Plan to Launch Censored Search Engine in China," press release, August 3, 2018.

34. Jack Poulson, "I Used to Work for Google. I Am a Conscientious Objector," *The New York Times,* April 23, 2019.

Chapter 13: A New World War

1. Rana Foroohar, "The Global Race for 5G Supremacy Is Not Yet Won," *Financial Times*, April 21, 2019.
2. Rana Foroohar, "'Patriotic Capitalism,'" *Financial Times*, October 8, 2018.
3. Rana Foroohar, "Globalised Business Is a US Security Issue," *Financial Times*, July 15, 2018.
4. Alliance for American Manufacturing, "American-Made National Security," press release.
5. U.S. Department of Defense, "Assessing and Strengthening the Manufacturing and Defense Industrial Base and Supply Chain Resiliency of the United States: Report to President Donald J. Trump by the Interagency Task Force in Fulfillment of Executive Order 13806," September 2018.
6. Michael Brown and Pavneet Singh, "China's Technology Transfer Strategy," GovExec.com, January 2018.
7. Daniel R. Coats, "Worldwide Threat Assessment of the U.S. Intelligence Community," Office of the Director of National Intelligence, 2019.
8. Rana Foroohar, "Government Contracts Become Amazon's New Target Market," *Financial Times*, May 26, 2019.
9. Louise Lucas and Emily Feng, "Inside China's Surveillance State," *Financial Times*, July 20, 2018.
10. Javier C. Hernandez, "Why China Silenced a Clickbait Queen in Its Battle for Information Control," *The New York Times*, March 16, 2019.
11. Adrian Shahbaz, "Fake News, Data Collection, and the Challenge to Democracy," Freedom House, 2018, https://freedomhouse.org/report/freedom-net/freedom-net-2018/rise-digital-authoritarianism.
12. Rana Foroohar, "China's Xi Jinping Is No Davos Man," *Financial Times*, January 20, 2019.
13. Ibid.
14. Jordan Robertson and Michael Riley, "The Big Hack: How China Used a Tiny Chip to Infiltrate U.S. Companies," *Bloomberg Businessweek*, October 4, 2018.
15. Lee outlines some of this in his book *AI Superpowers* as I do in *Makers and Takers*.
16. Rana Foroohar, "Advantage China in the Race to Control AI?" *Financial Times*, September 21, 2018.
17. Rana Foroohar, "Fight the FAANGs, Not China," *Financial Times*, May 6, 2018.
18. Foroohar, "China's Xi Jinping Is No Davos Man."
19. Lauren Easton, "How I Got That Photo of Zuckerberg's Notes," Associated Press, April 11, 2018.

20. Louise Lucas, "Huawei Deal with AT&T to Sell Phones in US Falls Through," *Financial Times,* January 8, 2019.

21. Mike Isaac and Cecilia Kang, "Facebook Expects to Be Fined Up to $5 Billion Over Privacy Issues," *The New York Times,* April 24, 2019.

22. David Shepardson, "Facebook Confirms Data Sharing with Chinese Companies," Reuters, June 5, 2018.

23. Rana Foroohar, "Facebook's Data Sharing Shows It Is Not a US Champion," *Financial Times,* June 6, 2018.

24. Luce and Rana, "Election Manipulation Edition."

25. Robert D. Atkinson and Michael Lind, "Who Wins After U.S. Antritrust Regulators Attack? China," *Fortune,* March 29, 2018.

26. Max Ehrenfreund, "A Majority of Millennials Now Reject Capitalism, Poll Shows," *The Washington Post,* April 26, 2016.

Chapter 14: How to Not Be Evil

1. Rana Foroohar, *Makers and Takers: How Wall Street Destroyed Main Street* (New York: Crown Business, 2016).

2. John Thornhill, "The Social Networks Are Publishers, Not Postmen," *Financial Times,* March 25, 2019.

3. Matt Novak, "New Zealand's Prime Minister Says Social Media Can't Be 'All Profit, No Responsibility,'" Gizmodo, March 3, 2019.

4. Open Markets Institute, "Key Judge Warns of Concentrated Power, Calls for Reviving Antitrust Tools," *Corner,* May 2, 2019.

5. Shapiro and Aneja, "Who Owns Americans' Personal Information and What Is It Worth?"

6. Ibid.

7. Cathy O'Neil, "Audit the Algorithms That Are Ruling Our Lives," *Financial Times,* July 30, 2018.

8. International Consortium of Investigative Journalists, "Explore the Panama Papers," January 31, 2017.

9. Author interviews with Joseph E. Stiglitz, 2017, 2018.

10. Rana Foroohar, "The Need for a Fair Means of Digital Taxation Increases," *Financial Times,* February 27, 2018.

Bibliography

Adams, Charles Francis Jr. *Railroads: Their Origins and Problems.* 1878.

Auletta, Ken. *Googled: The End of the World as We Know It.* New York: Penguin Books, 2009.

Battelle, John. *The Search: How Google and Its Rivals Rewrote the Rules of Business and Transformed Our Culture.* New York: Portfolio, 2005.

Bogost, Ian. *Persuasive Games: The Expressive Power of Videogames.* Cambridge, Mass.: MIT Press, 2010.

———. *Play Anything: The Pleasure of Limits, the Uses of Boredom, and the Secret of Games.* New York: Basic Books, 2016.

Brynjolfsson, Erik, and Andrew McAfee. *Machine Platform Crowd: Harnessing Our Digital Future.* New York: W.W. Norton, 2018.

Eckles, Dean, and B. J. Fogg, eds. *Mobile Persuasion: 20 Perspectives on the Future of Behavior Change.* Stanford, Calif.: Stanford Captology Media, 2007.

Edwards, Douglas. *I'm Feeling Lucky: The Confessions of Google Employee Number 59.* Boston: Houghton Mifflin Harcourt, 2011.

Eyal, Nir, with Ryan Hoover. *Hooked: How to Build Habit-Forming Products.* New York: Portfolio/Penguin, 2014.

Farrell, Joseph, Carl Shapiro, and Hal R. Varian. *The Economics of Information Technology.* Cambridge: Cambridge University Press, 2004.

Ferguson, Niall. *The Square and the Tower: Networks and Power, from the Freemasons to Facebook.* New York: Penguin Press, 2018.

Fisher, Adam. *Valley of Genius: The Uncensored History of Silicon Valley (as Told by the Hackers, Founders, and Freaks Who Made It Boom).* New York: Twelve, 2018.

Foer, Franklin. *World Without Mind: The Existential Threat of Big Tech.* New York: Penguin Press, 2017.

Fogg, B. J. *Persuasive Technology: Using Computers to Change What We Think and Do.* San Francisco, Calif.: Morgan Kaufmann, 2003.

Foroohar, Rana. *Makers and Takers: How Wall Street Destroyed Main Street.* New York: Crown Business, 2016.

Galloway, Scott. *The Four: Or, How to Build a Trillion-Dollar Company.* New York: Portfolio/Penguin, 2017.

Goldfarb, Brent, and David A. Kirsch. *Bubbles and Crashes: The Boom and Bust of Technological Innovation.* Stanford, Calif.: Stanford University Press, 2019.

Goldin, Claudia, and Lawrence F. Katz. *The Race Between Education and Technology.* Cambridge, Mass.: Belknap Press of Harvard University Press, 2008.

Gray, Mary L., and Siddharth Suri. *Ghost Work: How to Stop Silicon Valley from Building a New Global Underclass.* New York: Houghton Mifflin Harcourt, 2019.

Greene, Lucie. *Silicon States: The Power and Politics of Big Tech and What It Means for Our Future.* Berkeley, Calif.: Counterpoint, 2018.

Greenfield, Kent. *Corporations Are People Too (And They Should Act Like It).* New Haven, Conn.: Yale University Press, 2018.

Grunes, Allen P., and Maurice E. Stucke. *Big Data and Competition Policy.* Oxford: Oxford University Press, 2016.

Hartley, Scott. *The Fuzzy and the Techie: Why the Liberal Arts Will Rule the Digital World.* New York: Houghton Mifflin Harcourt, 2017.

Haskel, Jonathan, and Stian Westlake. *Capitalism Without Capital: The Rise of the Intangible Economy.* Princeton, N.J.: Princeton University Press, 2018.

Herman, Arthur. *Freedom's Forge: How American Business Produced Victory in World War II.* New York: Random House, 2012.

Isaacson, Walter, *Steve Jobs.* New York: Simon & Schuster, 2011.

Janeway, William H. *Doing Capitalism in the Innovation Economy.* Cambridge: Cambridge University Press, 2018.

Johnson, Nicholas L., and Alex Moazed. *Modern Monopolies: What It Takes to Dominate the 21st Century Economy.* New York: St. Martin's Press, 2016.

Kaye, David. *Speech Police: The Global Struggle to Govern the Internet.* New York: Columbia Global Reports, 2019.

Lanier, Jaron. *Who Owns the Future?* New York: Simon & Schuster, 2013.

———. *You Are Not a Gadget: A Manifesto.* New York: Random House, 2011.

Larson, Deborah Welch, and Alexei Shevchenko. *Quest for Status: Chinese and Russian Foreign Policy.* New Haven, Conn.: Yale University Press, 2019.

Lee, Kai-Fu. *AI Superpowers: China, Silicon Valley, and the New World Order.* New York: Houghton Mifflin Harcourt, 2018.

Levy, Steven. *In the Plex: How Google Thinks, Works, and Shapes Our Lives.* New York: Simon & Schuster, 2011.

Lynn, Barry C. *End of the Line: The Rise and Coming Fall of the Global Corporation.* New York: Doubleday, 2005.

Mayer-Schönberger, Viktor, and Kenneth Cukier. *Big Data: A Revolution That Will Transform How We Live, Work, and Think.* Boston: Houghton Mifflin Harcourt, 2013.

Mayer-Schönberger, Viktor, and Thomas Ramge. *Reinventing Capitalism in the Age of Big Data.* New York: Basic Books, 2018.

McNamee, Roger. *Zucked: Waking Up to the Facebook Catastrophe.* New York: Penguin, 2019.

Olson, Mancur. *The Rise and Decline of Nations.* New Haven, Conn.: Yale University Press, 1982.

O'Neil, Cathy. *Weapons of Math Destruction: How Big Data Increases Inequality and Threatens Democracy.* New York: Crown, 2016.

O'Reilly, Tim. *WTF? What's the Future and Why It's Up to Us.* New York: Harper Business, 2017.

Pasquale, Frank. *The Black Box Society: The Secret Algorithms That Control Money and Information.* Cambridge, Mass.: Harvard University Press, 2015.

Posner, Eric A., and E. Glen Weyl. *Radical Markets: Uprooting Capitalism and Democracy for a Just Society.* Princeton, N.J.: Princeton University Press, 2018.

Ramo, Joshua Cooper. *The Seventh Sense: Power, Fortune, and Survival in the Age of Networks.* New York: Little, Brown, 2016.

Rosenblat, Alex. *Uberland: How Algorithms Are Rewriting the Rules of Work.* Oakland: University of California Press, 2018.

Sandel, Michael. *What Money Can't Buy: The Moral Limits of Markets.* New York: Farrar, Straus and Giroux, 2012.

Schmidt, Eric, and Jared Cohen. *The New Digital Age: Transforming Nations, Businesses, and Our Lives.* New York: Vintage Books, 2014.

Shapiro, Carl, and Hal R. Varian. *Information Rules: A Strategic Guide to the Network Economy.* Boston: Harvard Business School Press, 1999.

Stone, Brad. *The Everything Store: Jeff Bezos and the Age of Amazon.* New York: Little, Brown, 2013.

Sundararajan, Arun. *The Sharing Economy: The End of Employment and the Rise of Crowd-Based Capitalism.* Cambridge, Mass.: MIT Press, 2016.

Taplin, Jonathan. *Move Fast and Break Things: How Facebook, Google, and Amazon Cornered Culture and Undermined Democracy.* New York: Little, Brown, 2017.

Tepper, Jonathan, with Denise Hearn. *The Myth of Capitalism: Monopolies and the Death of Competition.* Hoboken, N.J.: John Wiley and Sons, 2019.

Thompson, Clive. *Coders: The Making of a New Tribe and the Remaking of the World.* New York: Penguin, 2019.

Vise, David. *The Google Story: Inside the Hottest Business, Media, and Technology Success of Our Time.* New York: Bantam Dell, 2005.

Webb, Amy. *The Big Nine: How the Tech Titans and Their Thinking Machines Could Warp Humanity.* New York: Public Affairs, 2019.

West, Darrell M. *The Future of Work: Robots, AI, and Automation.* Washington, D.C.: Brookings Institution Press, 2018.

Wu, Tim. *The Attention Merchants: The Epic Scramble to Get Inside Our Heads.* New York: Knopf, 2016.

———. *The Curse of Bigness: Antitrust in the New Gilded Age.* New York: Columbia Global Reports, 2018.

Zuboff, Shoshana. *The Age of Surveillance Capitalism: The Fight for a Human Future at the New Frontier of Power.* New York: Public Affairs, 2019.

Index